PLANT MEMBRANES

PLANT MEMBRANES

A biophysical approach to structure, development and senescence

by

YA'ACOV Y. LESHEM

Department of Life Sciences, Bar Ilan University, Ramat Gan, Israel

with the participation of

ROBERT L. SHEWFELT

Department of Food Sciences and Technology, University of Georgia, U.S.A.

COLIN M. WILLMER

Department of Biological and Molecular Sciences, University of Stirling, U.K.

and

OMAR PANTOJA

Department of Botany, University of Toronto, Canada

Kluwer Academic Publishers

Dordrecht / Boston / London

Library of Congress Cataloging-in-Publication Data

Leshem, Ya'acov Y.
Plant membranes : a biophysical approach to structure, development and senescence / by Ya'acov Y. Leshem with the participation of Robert L. Shewfelt ... [et al.].
p. cm.
Includes bibliographical references and index.
ISBN 0-7923-1353-4 (acid free paper)
1. Plant membranes. I. Shewfelt, Robert L. II. Title.
QK725.L46 1991
581.87'5--dc20 91-23233

ISBN 0-7923-1353-4

Published by Kluwer Academic Publishers,
P.O. Box 17, 3300 AA Dordrecht, The Netherlands

Kluwer Academic Publishers incorporates
the publishing programmes of
D. Reidel, Martinus Nijhoff, Dr W. Junk and MTP Press.

Sold and distributed in the U.S.A. and Canada
by Kluwer Academic Publishers,
101 Philip Drive, Norwell, MA 02061, U.S.A.

In all other countries, sold and distributed
by Kluwer Academic Publishers Group,
P.O. Box 322, 3300 AH Dordrecht, The Netherlands.

Printed on acid-free paper

Printed in the Netherlands

Contents

Preface

The plasma membrane is at once the window through which the cell senses the environment and the portal through which the environment influences the structure and activities of the cell. Its importance in cellular physiology can thus hardly be overestimated, since constant flow of materials between cell and environment is essential to the well-being of any biological system. The nature of the materials moving into the cell is also critical, since some substances are required for maintenance and growth, while others, because of their toxicity, must either be rigorously excluded or permitted to enter only after chemical alteration. Such alteration frequently permits the compounds to be sequestered in special cellular compartments having different types of membranes. This type of homogeneity, plus the fact that the wear and tear of transmembrane molecular traffic compels the system to be constantly monitored and repaired, means that the membrane system of any organism must be both structurally complex and dynamic.

Membranes have been traditionally difficult to study because of their fragility and small diameter. In the last several decades, however, remarkable advances have been made because of techniques permitting the bulk isolation of membranes from homogenized cells. From such isolated membranes have come detailed physical and chemical analyses that have given us a detailed working model of membrane. We now can make intelligent guesses about the structural and functional interactions of membrane lipids, phospholipids, proteins, sterols and water. We can even simulate some aspects of membrane action in simulated constructs derived from membrane fragments or laboratory chemicals. Most importantly, we can now understand some aspects of membrane specificity in terms of immunologically identifiable surface receptors. These permit us to interpret the action of hormones, ions and other oligodynamic substances in the cell in terms of cascades of chemical events in the cell initiated by external stimuli.

The study of plant membranes has of course benefitted greatly by studies with animal and microbial cells, but the uniqueness of some cellular organelles, especially chloroplasts, has meant that plant scientists have had to break new ground. We have come to know, for example, that plant lipids differ from mammalian lipids in such details as glycerolipid acyl chain residues and types of glycerolipids. Similarly, phytosterols differ in some respects from those found in animals. These important differences in the chemical nature of major membrane components lead to noteworthy differences in membrane architecture and function.

This book represents a major new attempt to synthesize information about plant membrane chemistry and physics into an understanding of the significance of membrane properties in physiological processes. After a comprehensive review of necessary basic physical principles and of the chemical components of membranes and their interactions, it tries to depict how environmental stress, disease, senescence and growth can be understood in terms of changes in these parameters. It also includes special chapters on chilling-freezing injury and the unique membrane physiology of guard cells, written by guest authors. This information will be valuable to researchers, teachers and students alike. The author and his collaborators are to be congratulated for having produced a highly readable exposition of one of the most important subjects in all of biology.

Arthur W. Galston

Eaton Professor of Botany – Emeritus
Department of Biology
Yale University
New Haven, Connecticut

Acknowledgements

Much of this book's contents is the result of discussions, ideas and scientific interchange of ideas I have had during recent years with many friends and colleagues especially of two scientific organizations– the International Plant Lipid Biochemistry and Physiology Association and the International Plant Growth Substance Association who have encouraged me to produce this text. My students at the Bar Ilan University have added the leaven and incentive to undertake the task. Moreover workers in the field of plant adaptation to temperature stress and stomatal physiology have also provided important guidelines for my colleagues Drs. Robert Shewfelt, Colin Willmer and Omar Pantoja and myself in pertinent and related areas.

For initiating me, either by periods of co-operative research or by leisurely discussion, into the tortuous pathways of several aspects of membrane biophysics and biochemistry, I wish to offer special thanks to Prof. Meir Shinitzky of the Membrane Research Department, the Weizmann Institute of Science, Rehovoth, Israel, to Prof. John E. Thompson, Department of Biology, Waterloo University, Waterloo, Ontario, Canada and to Dr. Ian B. Ferguson of the DSIRO, Auckland, New Zealand.

Thanks for information and discussion of the assembled material on specific topics, not necessarily in the book chapter format, are also expressed to the following:– Prof. Ora Kedem, Department of Membrane Research, the Weizmann Institute, Israel, and Prof. Peter Steponkus, Department of Atmospheric Sciences, Cornell University, New York, U.S.A. for their remarks on membrane structure and water relationships; Dr. Ehud M. Landau, Department of Structural Chemistry, the Weizmann Institute for his introducing me to certain aspects of surface chemistry; Dr. Michael Edidin, Department of Biology, Johns Hopkins University, Baltimore, Maryland, U.S.A. for his guidelines on lateral diffusion in biomembranes; Prof. Norio Murata of the National Institute for Basic Biology, Myodaiji, Okazaki, Japan for his remarks on the role of lipid desaturation in temperature acclimation; Dr. Eva Selstam, of the Department of Plant Physiology, University of Umeå, Sweden for her information on cubic phases in plant membranes; Drs. Anna Stina Sandelius and Conny Liljenberg of the Botanical Institute, University of Göteborg, Sweden for their aid respectively on phosphatidylinositol metabolism and sterol involvement in membrane biophysics; Dr. Alistair Hetherington and Prof. Terry Mansfield of the Institute of Environmental and Biological Sciences, University of Lancaster, Lancaster, U.K. for their helpful and critical pointers on the behavior of stomatal guard cell membranes; and Prof. Steve Lindow of the Department of Plant Pathology, University of Berkeley, California, U.S.A. for novel information on the ice nucleating protein structure; Prof. John Goad, Department of Biochemistry, University of Liverpool, U.K. for his remarks on plant sterol metabolism; Dr. Ian Max Møller, Department of Plant Physiology, University of Lund, Sweden for his interesting insights on the structure and function of the plasmalemma; and finally Dr. Claude Penel of the Laboratory of Plant Physiology, University of Geneva, Switzerland for providing the opportunity to participate in a plant membrane workshop, conducted by plant membrane structure savants, where interchange of ideas and information were most helpful. Illustrative material was kindly supplied by several workers whose names and institutional affiliations are mentioned in the text.

I also wish to thank the U.S.A.–Israel Binational Research Foundation for a research grant (#86–002) which enabled execution of some of the experiments whose results are here reported, especially pertaining to Chapter 9 – "Membrane-associated Phospholytic and Lipolytic Enzymes."

I also wish to thank Dr. Yael Amzalak, Director of the Mishkenot Sha'ananim Center, Jerusalem and Mr. Teddy Kollek, Mayor of Jerusalem, for enabling me to utilize the residential writer's facilities of Jerusalem's Mishkenot Sha'ananim Quarter, the quiet and surroundings of which were an aid and inspiration for writing a considerable part of this text.

Thanks are also due to Ella Gindi of the Bar Ilan University's Technical office for her graphic work and to Sarah Lederhendler for her patient typing of the MS. For their painstaking reading and correction of the text I am deeply indebted to Philip Caplan of Asbury Park, New Jersey, and to Dr. Lee M. Spetner of Rehovoth. Last but not least I wish to thank Ir. Adriaan Plaizier, Scientific Editor of

Kluwer Publishers for his aid, encouragement and advice in producing this book.

Chapters 1–9 and Chapter 12 of this book were written by Y.Y.L., who also did the overall coordination and editing, Chapter 10 – "Responses of Plant Membranes to Chilling and Freezing" was written by Dr. Robert L. Shewfelt, Associate Professor, Department of Food Science and Technology, University of Georgia Agricultrual Experiment Station, Griffin, Georgia, U.S.A. and Chapter 11 – "The Tonoplast and Plasma Membrane of Guard Cells" jointly by Dr. Colin M. Willmer, Senior Lecturer, Department of Biological and Molecular Sciences, University of Stirling, Scotland, U.K. and by Dr. Omar Pantoja, Research Associate, Department of Botany, University of Toronto, Canada.

Y. Y. L.
Ramat Gan and Jerusalem

Tevet, 5752 – December 1991

1. Introduction

Research at a basic level has indicated that in many respects plant membranes resemble those of mammals or of bacteria but that nevertheless certain basic differences exist. Moreover, plants possess a specific organelle, the chloroplast, which is lacking in mammals. An often overlooked fact is that due to the preponderance of plants and green algae, the chloroplast membrane system is the most ubiquitous one in nature.

1.1. Plant membrane categories

As will be described in greater detail in later chapters plant membrane categories include:

1. The *plasmalemma* or plasma membrane. Steer (1988) has advanced an interesting hypothesis that this membrane undergoes *complete turnover* in a brief period varying from 10 minutes to 2 hours and that in this process components of other membrane categories listed below also participate. This concept is not without its contenders but be it as it may, it is now of general acceptance that in membrane adaptation to environmental change (heat or cold) or to stress, *retailoring* occurs – in this process only certain parts of the plasmalemma, e.g. fatty acyl tailgroups, are exchanged. An up-to-date description of various aspects of plasmalemmal physiology is to be found in the book edited by Larsson and Møller (1990).
2. The *tonoplast* – this being the membrane which encompasses the vacuole which has been termed by Matile (1990) "the toxic compartment" of plant cells. There is partial evidence that the tonoplast resembles the plasmalemma in its composition (Canut et al., 1991).
3. *Plastid membranes.* These include membranes of the various cell-

contained plastids and particularly the chloroplast envelope and thylakoid membranes.
4. *Mitochondrial membranes.*
5. The *endoplasmatic reticulum* (ER) membrane system which is highly folded.
6. *Stomatal guard cell membranes.*
7. *Miscellaneous* – inclusive of the membranes surrounding the nucleus, lysosomes, fatty bodies, peroxisomes and Golgi apparatus. During the last decade plant research in this field has primarily dealt with categories 1–6.

1.2. The amphiphilic nature of biomembranes

According to current and contemporary concepts (Evans, 1988) the basis of membrane structure lies in its amphiphilic (or amphipathic) nature. In 1936, Hartley introduced the term *amphipathy* to describe the unusual properties of aqueous solutions of detergent molecules:

> This unsymmetrical duality of affinity is so fundamental a property of paraffin-chain ions, being directly responsible for all the major peculiarities of paraffin-chain salts in aqueous solutions, that it is worth while to give it a special name. The property is essentially the simultaneous presence (in the same molecule) of separately satisfiable *sympathy* and *antipathy* for water. I propose, therefore, to call this property – *amphipathy* – the possession of both feelings.

By biophysical definition an *amphiphile* is a molecule or a structure which possesses both hydrophilic and hydrophobic sections. Amphiphiles self-assemble into a variety of microstructures that are *physically, not chemically*, associated. These structures, which include micelles, vesicles, liposomes, microtubules, and bilayers, constitute microphases possessing oil-like regions and large interfacial areas. Many biological and industrial processes utilize their properties.

In biological processes, amphiphiles continuously transform from one microstructure to another in response to delicate changes in concentration, pH, ionic strength, and temperature. *In conjunction with proteins, sterols and carbohydrates, they form the super assemblies which mediate and control life processes.*

In industrial applications, such as the utilization of amphiphiles as drug delivery systems or microdomains for synthesis of small metal clusters, the goal is to set structure, maintain its integrity under adverse conditions, and then transform or disperse it at the end of the process. Successful realization of these goals depends on the ability to dictate the assembly, stability and reactivity of the microstructures.

Phospholipids, *galactolipids* and *sulpholipids* are three predominant types of plant membrane-contained amphiphiles. Since all three are derived from glycerol, the designation *glycerolipids* is often employed as a common denominator. *Sphingolipids*, also amphiphiles, are the second largest class of membrane lipids. In this lipid group the basic alcohol is not glycerol but *sphingenine* or one of its derivatives.

1.3. Self-assembly to form a basic cell or organelle encasement

When randomly introduced into an aqueous environment, *phosphatidylcholine*, the most abundant plant and animal phospholipid membrane component and some but not all (cf. Chapter 5) of the other glycerolipid molecules undergo rearrangement in which the individual molecules spontaneously arrange themselves into a bilayer. In the bilayer the molecules in both layers align themselves in such a manner that their longitudinal axes are roughly perpendicular to the plane of the bilayer. The hydrophilic headgroups face water on both sides of the bilayer, and the hydropholic tails position themselves in the middle of the bilayer thereby expelling water from it. This arrangement is the state of lowest free energy for these molecules in an aqueous medium, as is the essential spherical shape that such a super-assembly assumes.

In simplistic terms, such amphiphiles spontaneously form a closed envelope with considerable mechanical strength, since if any edge of the bilayer were in contact with water, hydrophobic tails along the edge would be exposed; hence the bilayer closes to form a vesicle effectively segregating the fluid inside the vesicle from the fluid surrounding it. The conceptual analogy to the biological unit cell is evident.

Notwithstanding the exclusion of water from the internal hydrophobic section of the bilayer occupied by the water repellant fatty acyl chains of the glycerolipids, a certain degree of transmembrane water exchange does occur; this is believedly due to the existence of ultrastructural longitudinal microconduits parallel to the longitudinal axes of the glycerolipid fatty acyl tails or even to the presence of water channels (see Chapter 6).

Concerning the second membrane component – the proteins – it is commonly held that their mobility is dependent upon the physicochemical state of the lipid matrix in which they float (Carruthers and

Melchior, 1986). The third group of membrane constituents – the sterols – as explained later in detail in Chapter 4 – serve as stabilizing agents on membrane fluidity acting in concert with the lipid components. Various categories of plant membranes contain different relative amounts of sterols (and while these are integral components of the plasmalemma, mitochondrial membranes of many species contain very little sterol, if any). However, according to Rochester et al. (1987), as a rule, sterol composition is not a characteristic of a certain membrane but rather of a certain organ or of environmental factors.

The spontaneous tendency of the amphiphilic membrane components to form closed vesicles, for some as yet unknown reason, in monocots is more pronounced than in dicots. This phenomenon has pertinence in plant cell culture and genetic engineering where the incorporation *Agrobacterium tumifasciens* vector-borne DNA sections bearing specific information, e.g. herbicide resistance, encounters difficulties due to the rapid self-sealing off of any microlocus of protoplast* membrane penetration of the transgenic DNA. In corn (*Zea mays*) Gordon-Kamm et al. (1990) have overcome this difficulty by electric microprojectile bombardment with tungsten particles coated with the selectable marker gene thus causing self-sealing holes – the microlesion thus produced being markedly less prone to premature sealing off. In this manner the above workers have produced a transgenic corn cultivar resistant to the herbicides bialaphos, which contains the herbicidal compound phosphinothricin, and gluphosinate.

1.4. Liposomes

An artefactal analogy to a biological vesicle is termed a *liposome*. In pharmaceutical science and also to an as yet limited extent in plant genetic engineering the use of liposomes – which essentially are lipid sacs housing aqueous content – for drug or informational delivery, in specific cases is a more effective means than standard administration. Such delivery may take place at specific sites where the "carrier" liposome sheath merges with the membrane bilayer of the treated

**protoplast* – this is a term, employed in plant tissue culture, to describe a whole intact cell deprived of encasing cell walls by pretreatment with cell wall-digesting enzymes. In reality the designation "protoplast" is a misnomer since in addition to protoplasma, the denuded cell produced is surrounded by a plasma membrane and retains all of its contents including organelles and nucleus.

tissue, so doing emptying its pharmaceutical aqueous content in a very localized and specific manner.

Liposome-mediated virus transfection and gene expression is well documented for plant protoplasts. However, the use of liposomes is not restricted to protoplasts. There is evidence that liposomes fuse with or enter into germinating pollen (Gad et al., 1988). There is some success in insertion of DNA into callus (Rosenberg et al., 1990).

Plant research concentrating on genetic engineering and gene delivery as reviewed by Altman's group (cf. Gad et al., 1990) indicates that liposomes have been used extensively to entrap a large variety of molecules, including macromolecules such as enzymes and nucleic acids, and to deliver them into the cells by fusion with the plasma membrane or through endocytosis. Possible advantages of the use of liposomes for gene transfer include: (1) enhanced delivery of the sequestered DNA (or RNA), (2) protection of entrapped DNA from nuclease activity, either in the culture medium, or in the tissue, or both, (3) targeting to specific cells and improved interaction by modified liposomes, (4) delivery of nucleic acids into a variety of cells besides protoplasts, through cell wall, by entry through plasmodesmata, (5) delivery of intact small organelles (e.g. mitochondria) and (6) low toxicity to cells.

Three major types of liposomes are employed:

(1) SUV, small unilamellar vesicles [diameter 30–50 nm, volume <0.21].
(2) LUV, large unilamellar vesicles [diameter >150 nm, volume >21].
(3) MLV, large multilamellar vesicles [diameter >500 nm, volume >21].

LUV are the best choice for routine delivery. Although the preparation of LUV is somewhat more tedious than that of the other two types of liposomes, the benefits are noteworthy: (1) the encapsulation volume is appropriate for entrapping most of the macromolecules in use, (2) an extremely high degree of reproducibility is possible when following exactly the already existing detailed protocol, (3) a homogeneous sizewise population following extrusion through polycarbonate filters.

The major drawback of liposome-mediated gene expression is its low efficiency. The reason for this is not clear, but one likely reason is related to the lipsome/cell ratio. In most cases where the cells were literally flooded with liposomes, rough estimates point to a ratio of more than 10^5 liposomes per tobacco protoplast (Rosenberg et al., 1990).

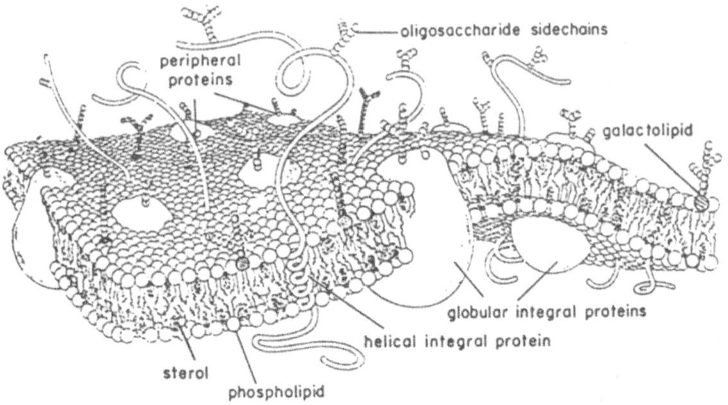

Fig. 1.1. A typical bilayered biological membrane (after Braetscher, 1985). With regard to the membrane proteins this is sometimes referred to as the "apple and pear" model.

1.5. Layering of plant membranes

The term *bilayer* describes the most prevalent layering type of plant membranes. Each monolayer of the bilayer is termed a *leaflet*. Layered formations in plant membrane may be classified under three groups (see Fig. 1.1).

(1) *Single bilayer* – this formation which is the most common one is encountered in the plasmalemma, tonoplast and thylakoid membranes in chloroplasts and in the ER.
(2) *Double bilayer* – double bilayers are typical of the chloroplast envelope (see Fig. 1.2 and also Fig. 7.2 in Chapter 7) and of the mitochondrial membrane.
(3) "*Half-unit membranes*" – this interesting but rare configuration is comprised of only *one* leaflet i.e. one monolayer, and forms the membrane of *oleosomes* and *spherosomes* in triglyceride-rich fatty oil-containing seeds (Yatsu and Jack, 1972). In such half-unit membranes, the fatty phospholipid fatty acyl tails face inwards towards the content of the organelle due to the hydrophobic interaction between these tails and the tri-glyceride content of the lipid bodies (Huang et al., 1987). While only limited information exists on half-unit membranes, as presently documented it appears that they are found mainly in seeds. Extensive electron microscope-aided research for similar structures in fatty fruits such as *Persea americana* (avocado) conducted by W. Thompson at the University of California has not revealed their presence. An interesting object for research in this respect would be the olive fruit.

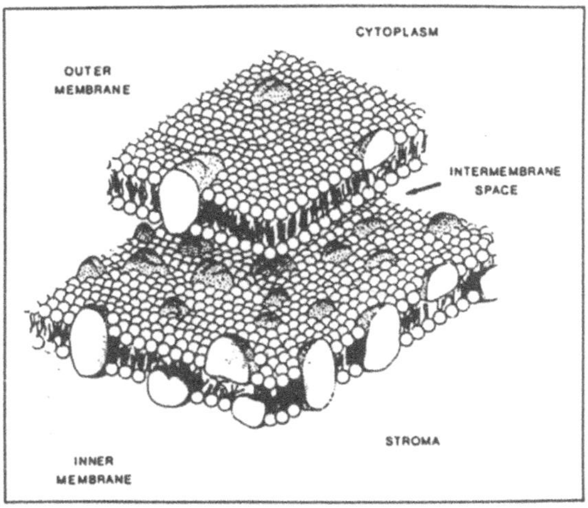

Fig. 1.2. The double bilayer comprising the chloroplast envelope (after Douce and Joyard, 1979).

We here point out that the above layered configurations are by no means the only membrane formations found in the plant membranes and that in response to adverse environmental conditions or as a function of specific glycerolipid composition, the bilayer or local domains in the bilayer may adopt non-bilayer configurations.

Non-bilayered biophysical membrane phases cannot be major structural elements in biological membranes since these would not form a continuous barrier surrounding the cell contents. However conditions leading to the formation of non-bilayer phases alter the properties of membranes while still in the bilayer phase. This topic will be dealt with in Chapter 7.

1.6. References and recommended reading

Bretscher, M.S. 1985. The molecules of the cell membrane. *Sci. Amer.* 253:86–90.

Canut, H., Justin, A.M., Demandre, C., Boudet, A. and Mazliak, P. 1991. Molecular species of PC and PE in plasmalemma and tonoplast isolated from soybean hypocotyls by free flow electrophoresis. pp. 23–26. In: Eds. P. J. Quinn and J. L. Harwood, *Plant Lipid Biochemistry – Structure, Utilization and Function*. Portland Press, London and Colchester.

Carruthers, A., and Melchior, D.L. 1986. How bilayer lipids affect membrane protein activity. *Trends Biochem. Sci.* 11:331–35.

Douce, R. and Joyard, J. 1979. The chloroplast envelope: an unusual cell membrane system. pp. 79–98. In: Eds. L.A. Appelqvist and C. Liljenberg. *Advances in the Biochemistry and Physiology of Plant Lipids*. Elsevier Press, Amsterdam.

Evans, D.F. 1988. *Self organization of amphiphiles*. Langmuir 4:3–12.

Gad, A.E., Rosenberg, N. and Altman, A. 1990. Liposome-mediated gene delivery into plant cells. *Physiol. Plant.* 79:177–83.

Gad, A., Zeewi, B. and Altman, A. 1988. Fusion of germinating watermelon pollen tubes with liposomes. *Plant Sci.* 55:69–75.

Gennis, R.B. 1989. *Biomembranes: Molecular Structure and Function.* Springer Verlag, Berlin.

Gordon-Kamm, W.J., Spencer, M., Mangano, M.L., Adams, T.R., Daines, R.J., Start, W.G., O'Brien, J.V., Chambers, S.A., Adams, W.R., Willets, N.G., Rice, T.B., Mackey, C.J., Krueger, R.W., Kausch, A.P. and Lemaux, P.G. 1990. Transformation of maize cells and regeneration from fertile plants. *Plant Cell* 2:603–18.

Huang, H.C., Qu, R., Wang, S., Vance, V., Cao, Y. and Lin, Y. 1987. Synthesis and degradation of lipid bodies in the scutella of maize. pp. 239–46. In: Eds. P. K. Stumpf, J. D. Mudd and W. D. Nes. *The Metabolism, Structure and Function of Plant Lipids.* Plenum Press, New York.

Larsson, C. and Møller, I.M. 1990. *The Plant Plasma Membrane.* Springer Verlag, Berlin.

Matile, P. 1990. The toxic compartment of plant cells. pp. 657–68. In: Eds. H.J.J. Nijkamp, L.H.W. van der Plas and J. van Aartrijk. *Progress in Plant Cellular and Molecular Biology.* Kluwer Academic Publishers, Dordrecht.

Quinn, P.J. 1976. *The Molecular Biology of Cell Membranes.* University Park Press, London.

Rochester, C.P., Kjellbom, P. and Larsson C. 1987. Lipid composition of plasmamembranes from barley leaves and roots, spinach leaves and cauliflower inflorescences. *Physiol. Plant.* 71:257–63.

Rosenberg, N., Dekel-Reichenbach, M., Navot, N., Gad, A.E., Altman, A. and Czosnek, H. 1990. Liposome-mediated introduction of DNA into plant protoplasts and calli. *Acta Hortic.* 280:509–16.

Steer, M.W. 1988. Plasma membrane turnover in plant cells. *Jour. Exp. Bot.* 39:987–96.

Sussman, M.R. and Harper, J.F. 1989. Molecular biology of the plasma membrane of higher plants. *Plant Cell* 1:953–60.

Yatsu, L.Y. and Jack, T.J. 1972, Spherosome membranes: half unit membranes. *Plant Physiol.* 49:937–43.

Part one

The building blocks of the membrane

2. Membrane fatty acyl tailgroups

2.1. Glycerolipid fatty acyl chains

As a background setting for this chapter we point out that the typical glycerolipid amphiphiles of the plant membrane are methyl esters of glycerol in which the carbon atoms nos. 1 and 2 – designated *sn*-1 and *sn*-2 (*sn* = specific numeration) are attached to long chained fatty acids. With certain exceptions, in most phospholipids and galactolipids in both plant and animal membranes the *sn*-1 site is usually occupied by a saturated fatty acid, while the *sn*-2 site contains a polyunsaturated fatty acid. This is exemplified by the scheme indicated in Fig. 2.1.

However in membrane-contained galactolipids especially those found in chloroplast membranes, a situation may occur in which *both* fatty acyl chains may be unsaturated or the order – *sn*-1 saturated and *sn*-2 unsaturated – reversed (See Fig. 3.9 in Chapter 3). Other biologically rare but interesting exceptions are neural membranes of the *Torpedo* (electric eel) in which both sites contain polyunsaturated fatty acyl residues and, as in a converse situation (where both tailgroups contain saturated fatty acids) in dipalmitoylphosphatidylcholine (DPPC) a minor phospholipid component which is found in *Vicia faba* and *Commelina commumis* stomatal guard cell microsomal membranes (Leshem et al., 1990), this DPPC also comprising the major pulmonary phospholipid in lung surfactant (Harwood, 1987).

Meticulous experimentation by Norio Murata's group in Okazaki, Japan (Wada et al., 1990; Tasaka et al., 1990) has demonstrated the presence of a further type of "double saturation", in the form of dipalmitoylphosphatidylglycerol (DPPG), present in chloroplast membranes of both prokaryotic and eukaryotic plants and which as discussed in detail in Chapter 10 may be involved with cold acclimation. It is moreover of particular interest that in this chloroplast-specific species the *sn*-2 chain instead of containing the saturated 16:0 species,

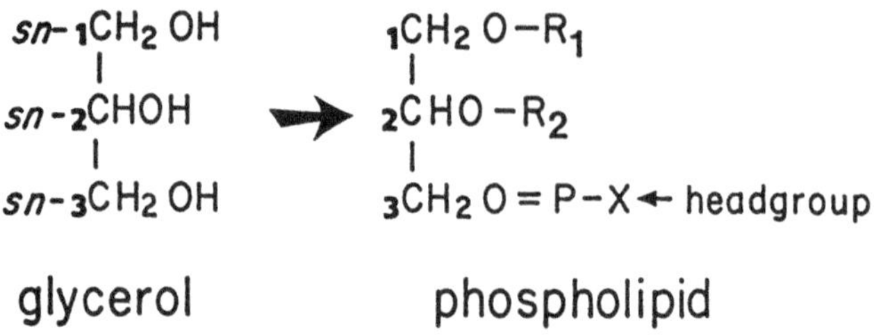

Fig. 2.1. Overall glycerol derivation of membrane phospholipids.

may sometimes contain a *trans*-monounsaturated 16:1 species. The latter, because of its *trans* double bonding, lacks the typical "kinking" of its *cis*-mono-unsaturated counterpart and in this respect structurally and biophysically has much in common with the saturated 16:0 species.

2.2. Sphingolipid fatty acyl chains

The topic of membrane sphingolipids in greater detail is dealt with on pg. 40 in Chapter 3 and here we limit discussion to the nature of the *single* fatty acid bound by means of an amine bond to the sphingosine alcohol moiety in glucosylceramide (see Figs 2.2, and Fig. 3.8 in the next chapter) which is the predominant sphingolipid in plants. It is here of interest to point out that the fatty acyl residues which to date have been characterized in plants and animals are long-chained saturated species with small amounts of monosaturated species. What recently has come to light is that fatty acyl residues in plant cerebrosides are usually 22:0, 24:0, 22:1 and 24:1 species. Animal sphingolipids in addition contain shorter saturated fatty acid residues including 16:0 and 18:0, and also 18:1.

The essentially saturated nature of the sphingolipid group of membranes, coupled to the saturated nature of the alcohol sphingosine (its

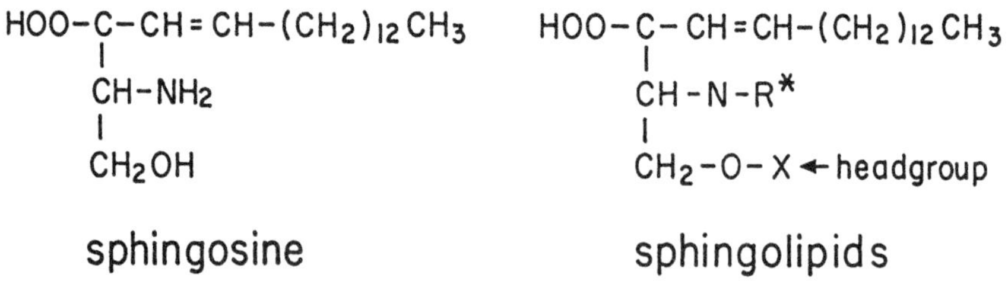

Fig. 2.2. The alcohol sphingosine and not glycerol is the basic building block for membrane sphingolipids. *R** is usually a long chained saturated fatty acid or a monounsaturated one.

single double bond has a *trans* configuration and concomitantly behaves biophysically as a saturated residue) causes this group of compounds as a whole to exert a rigidifying effect on the membrane, this being secondary to that of membrane sterols.

Since later chapters of this book, delving deeper into biophysical and biochemical regulatory mechanisms, employ terms and concepts which in the lipid literature are not always uniform and thus often not completely clear to many a scientist who is not a "lipidologist", the following section will outline some standardly accepted overall principles of biological fatty acyl composition and nomenclature. An overall grasp of these concepts is essential for further reasoning and representation as this text unfolds:

2.3. Some rules of membrane fatty acyl structure

1. The number of C atoms, indicating chain length is invariably even and varies from 12–26. However there have been reports of bacterial fatty acids with an uneven number of C atoms. The accepted reason for the even number is that these acids are derivatives of acetyl coenzyme A which itself has an even number of C atoms.
2. If a double bond (C=C) occurs, it is usually, but not always, in the middle of the chain e.g. in oleic acid (18:1) – between carbons # 9 and 10. According to present opinion the unsaturation is inserted as a secondary process into saturated chains by the *desaturase* group of enzymes.
3. If more than one double bond occurs, the distance between one double bond and the next is always 3 carbon atoms. In most organisms, especially in higher plants and mammals, methylene interruption between double bonds is maintained and conjugated double bonds are extremely rare. Thus chain elongation usually alternates with desaturation (i.e. desaturase activity) to maintain methylene interruption in polyunsaturated chains.
4. The double bond usually assumes a *cis* orientation. (A notable exception to this rule is *eliadic acid* which like *oleic* acid is 18:1, but differs from the latter in that its double bond has a *trans* orientation. See Table 2.1). From a stereostructural point of view it is stressed that this *cis* orientation induces a bend or a "kink" of approximately 44° in the chain, while the homologous *trans* acid is straight (See Fig. 2.3). This implies that if we have two fatty acids each having the *same* number of C atoms, one being saturated and the other unsaturated, due to the "kink" the *unsaturated one*

Table 2.1. Physical dimensions of fatty acids prevalent in plant membranes (after Cook, 1985)

Fatty acid	Abbreviation	Length (nm)	Molecular area (nm^2)	Configuration
Stearic	18:0	2.1	0.20	
Oleic	18:1	1.6	0.48	
Eliadic	18:1*	1.9	0.23	
Linoleic	18:2	1.3	0.82	

* This rare fatty acid has a *trans* C=C bond.

will be shorter than the saturated one. Moreover since these entities in a biological state are not stationary and possess, inter alia, rotationary motion (See Plate 2.1) it is apparent that the surface area covered by an unsaturated fatty acid will be greater than the surface area covered by a saturated one. These two parameters – fatty acyl length and surface area – play an important role in determining the properties of a biological membrane. Table 2.1 illustrates the above principle.

2.4. Biochemical nomenclature of naturally occurring fatty acids

Due to their importance as membrane components of "neutral fats" (tri-, di-, or mono-glycerides which are not integral membrane components but rather metabolic storage reserves in the form of plant oils and fats, waxes, etc.), this section will outline in detail, the accepted nomenclature employed to describe naturally occurring fatty acids.

At the outset two important points should be made. The first, as indicated by Stymne et al. (1991), is that while the various types of plant triacylglycerols have a choice of several scores or even hundreds of various types of fatty acyl residues, both saturated and unsaturated, in most plant membrane glycerolipids only eight main fatty acyl chains are utilized. This principle is illustrated in Fig. 2.4. The second point is that barring advanced stages of plant senescence, stress or pathological infection, very little fatty acid exists in the free state. Free fatty acids which may appear after hydrolytic action of senescence-associated phospholipases act as detergents in the cell and have a lytic and disruptive action on probably all the various categories of plant membranes and hence are fraught with destruction.

From a medicinal viewpoint polyunsaturated fatty acids are of importance since when present in the bloodstream they may prevent or slow down deposition of lipid material such as cholesterol in the

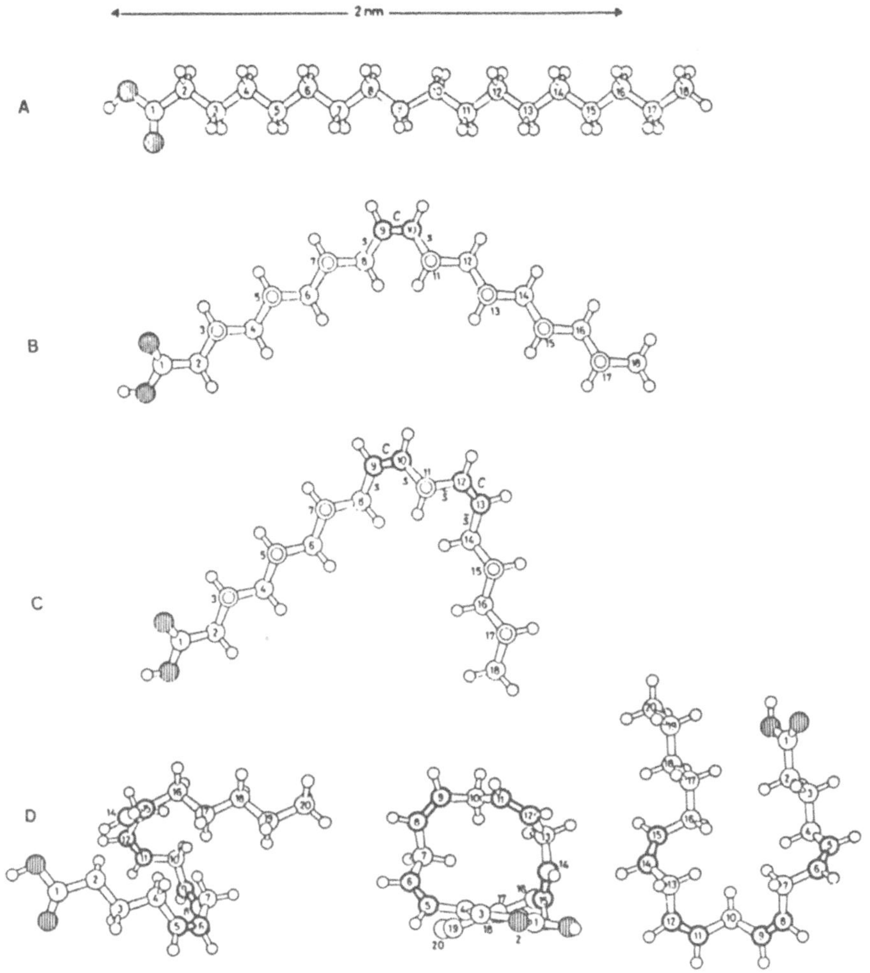

Fig. 2.3. Some of the commonly occurring fatty acids in the plant membrane (After de Jong, 1980). The *cis* unsaturated C═C double bonds are drawn with thicker lines. Single bonds have a length of 0.154 nm and an angle of approximately 111°; double bonds have a length of 0.133 nm and an angle of approximately 123°. A = stearic acid (18:0); B = oleic acid (18:1); C = linoleic acid (18:2); D = arachidonic acid (20:4). In this line the diagrams on the left and in the middle are respectively a side view and a top view of the helical form which is believed to be the most common biological configuration of arachidonic acid. The diagram on the right is the "U" form currently considered as being less probable.

interior of blood vessels and so doing alleviate cardial complications and atherosclerosis. In this respect there is considerable current interest in polyunsaturated acids which contain three or more double bonds. A similar lytic action on vascular cholesterol-deposit is attri-

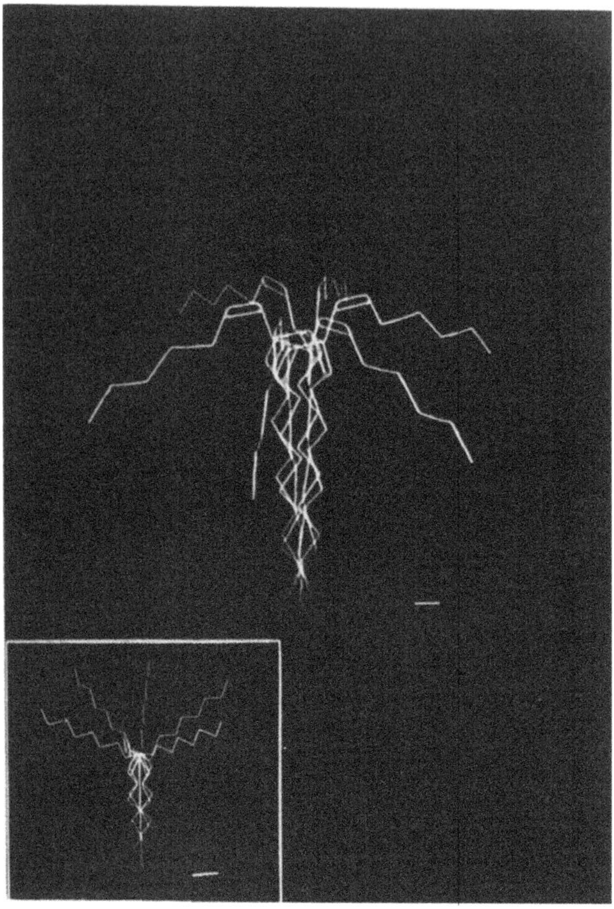

Plate 2.1. A stroboscopic photograph of rotational motion of linoleic acid (18:2) and of oleic acid (18:1 – insert). Note that because of the "kinks" (C═C bonds) linoleic acid describes a greater circle than does oleic acid, but that it is shorter than the latter. Scale line = 0.1 nm.

buted to a particular type of linolenic acid – the γ type which is present in seeds of the Evening Primrose (*Oenothera drummondi*) and in the Borage (*Borula*) plant. The polyunsaturated acids belonging to the omega (ω)-3 or *n*-3 (detailed later) category believed to be effective in this respect, include eicosapentaenoic acid (EPA) and docosahexaenoic acid (DHA) which respectively contain 5 and 6 double bonds, these being 20:5 and 22:6 species (*eicosa* = 20, and *docosa* = 22 in Greek, while the suffix *-ene* designates a C═C bond).

While these very long chained polyunsaturated fatty acids are found in higher plants and in algae, according to the present state of our

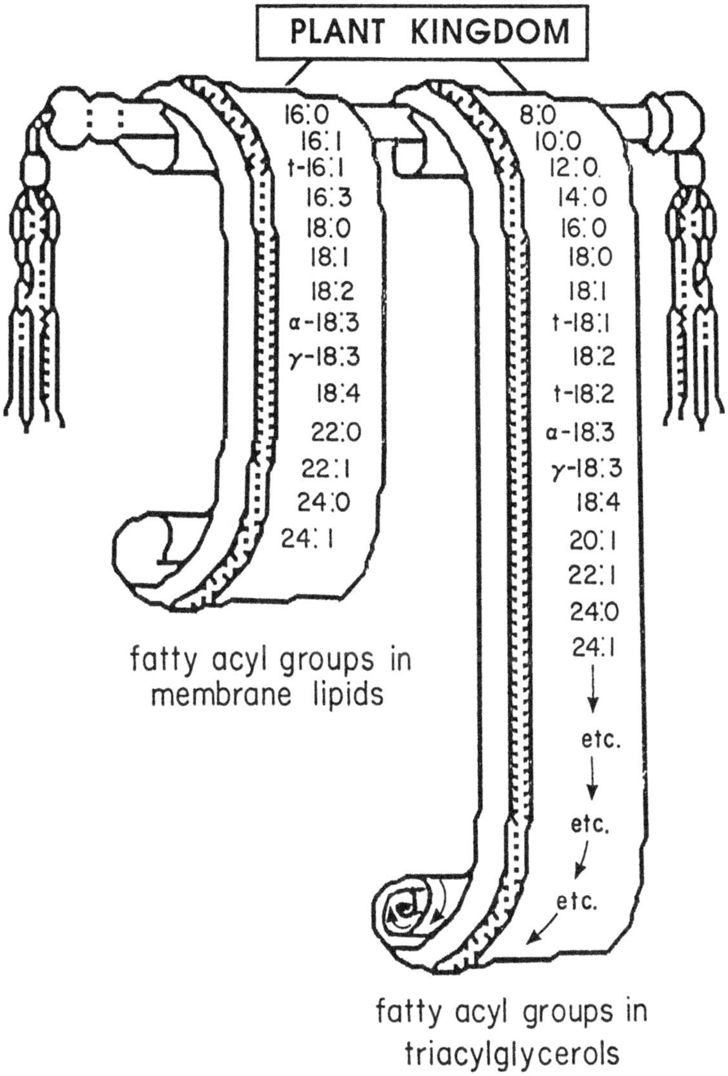

Fig. 2.4. Only relatively few fatty acyl residues are present in plant membrane glycerolipids: however several scores of widely diverse fatty acids are found in plant triacylglycerols. (Modified from a diagram provided with the courtesy of Dr. S. Stymne, University of Lund, Sweden).

knowledge, the occurrence in the Plant Kingdom of mammalian eicosanoid analogs such as prostaglandins, thromboxanes, leukotrienes and lipoxins is less clear. Tarchevsky and Grechkin (1989) of the Kazan Institute of Biology in the U.S.S.R. have critically studied the issue and have concluded that most of the evidence hitherto affirming the occurrence of such eicosanoid analogs in plants is tenuous. However they do concede that oxygenated metabolites of 18:3 ω^6 and 18:2

ω^6 fatty acids (the detailed nomenclature of which will presently be discussed) may be considered as valid eicosanoid analogs due to their similarity of biogenesis and regulatory activity. The common trivial name proposed by them for this category of plant compounds is "*linolenoids*." These include plant senescence-associated *jasmonic* and *traumatic* acids.

Sanders (1990) has pointed out essential and basic differences between dietary effects of linoleic and linolenic acids on the one hand, and EPA and DHA on the other. While relatively low intake of linoleic acid (ca 10% of energy intake) indeed has beneficial effects and promotes formation of prostacyclin, a prostaglandin metabolite that protects the blood vessel wall, high intakes of the same acid suppress the immune system and are associated with increased risk of formation of gallstones. High intake of linoleic acid also has the disadvantage of suppressing the metabolism of the "*n*-3" family of fatty acids (detailed in the next subsection in this chapter) including EPA and DHA which as mentioned above have amelioratory effects on coronary heart disease symptoms.

EPA and DHA are found in fish oil and in some species of algae, and in far lower amounts in some meats. *Preformed* EPA and DHA are rapidly incorporated into membrane lipids. Linolenic acid is converted to EPA but is far less potent in raising membrane lipid levels than preformed EPA. DHA too can be endogenously synthesized from linolenic acid, but rate of synthesis is very low and humans may require small amounts of preformed DHA in their diets (Sanders, 1988).

Many carnivorous mammals if deprived of eicosanoid-rich food such as certain varieties of fish and meat, never-the-less possess the ability to synthesize these eicosanoids from 18:2 or 18:3 of plant origin. This option apparently exists in vegetarians. However, individuals or population groups may vary in this respect. A group *not* possessing this ability is the Greenland Eskimos who have been found to be "obligate predators".

Extensive surveys of human populations have indicated maximal life expectancy in three populations – of Hokaido (the northern island of Japan), of Iceland and of Greenland Eskimos – this being approximately 78 years. Caroll (1986) has investigated the dietary habits of these populations and reported that a high consumption of fish is a common dietary feature. In laboratory tests performed on rats maintained on EPA and DHA diets, in addition to an increasing effect on longevity and a reduction of cholesterol-related heart disorders, incidence of intestinal and prostrate tumors was also markedly reduced.

In outlining the following terms, the generally accepted International Union of Biochemists' recommended system is employed (cf. Cook, 1985):

1. To indicate that oleic acid is an 18-carbon fatty acid with one double bond, the shorthand 18:1 is used. The number before the colon denotes the number of carbon atoms, and the number following refers to the number of double bonds. Linoleic acid for example is abbreviated 18:2 and linolenic acid 18:3, respectively possessing 2 and 3 double bonds and being 18 carbons long.
2. The position of an individual double bond or the specificity of an enzyme inserting it is indicated by the Greek letter delta (Δ). This describes the bond location relative to the *carboxyl* (COOH) end of the fatty acid chain. Thus the common isomer form of linolenic acid in more detail is designated 18:3 $\Delta^{9,12,15}$. In membrane glycerolipids the carboxyl ends are esterified to the 1st (*sn*-1) and 2nd (*sn*-2) hydroxyl (OH) of the glycerol backbone.

   The Δ^9 is a commonly occurring site of initial unsaturation of glycerolipid membrane unsaturated tails but exceptions, pertaining to triacylglycerols as well, do occur. These include the γ isomer of linolenic acid in which the initial C=C bond is at C-carbon atom no. 6 and is therefore an 18:3 $\Delta^{6,9,12}$ species.
3. To indicate the geometric configuration of a double bond, the designation is preceded by a *c*- for *cis* or *t*- for *trans*. Thus, *c*-18:1 distinguishes oleic acid from its *trans* isomer, eliadic acid. In plants and in higher animals the double bond is usually *cis*. An interesting case of occurrence of a fatty acid containing a *trans* configuration is found in the chloroplast thylakoid membrane reported by Trémolières et al. (1982) where a hexadecanoic acid *t*-16:1, *t*-Δ^3 in the fatty acyl tail of phosphatidylglycerol (see next Chapter) is believed to fulfill a light harvesting function in the process of photosynthesis. According to these workers, this particular type of membrane-associated fatty acyl residue is present only in higher plants and is absent from blue-green algae and from photosynthetic bacteria.
4. Amongst other conventions, one receiving increasing mention is to add further information as to the location of the double bond, one system of which indicates this not from the carboxyl tail, but from the free methyl endgroup. In this system of nomenclature the "*n*" terminology is employed (which is analogous to the less common "omega"–ω–designation). To give an example of application of this system, linoleic acid, mentioned above, would be represented as follows: 18:2 (*n*-6) $\Delta^{9,12}$. Due to pertinence of this parameter to regulatory membrane mechanisms and to human diet and health they have been grouped into "families" as follows:

The *n-9* group:	oleic acid – 18:1.
The *n-7* group:	palmitooleic acid – 16:1. Very rare, appearing in some geophytes e.g. Hellebore.
The *n-6* group:	linoleic acid – 18:2.
	γ linolenic acid – 18:3. In *Oenothera* and *Borago*.
	arachidonic acid – 20:4.
The *n-3* group*:	α linolenic acid – 18:3
	eicosapentaenoic acid (EPA) – 20:5
	docosapentaenoic acid (DPA) – 22:5
	docosahexaenoic acid (DHA) – 22:6

(*often termed ω-3, especially in commercial preparations)

Table 2.2 presents nomenclature, sites and type of double bonding of the major long chained fatty acids occurring in nature.

2.5. Differences between plants and mammals in membrane glycerolipid fatty acyl chain residues

2.5.a. *At the* sn-*1 site*

In membrane glycerolipids both in plants and animals, the *sn*-1 site is usually occupied by a saturated and therefore a straight-chained fatty acid this being either *palmitic acid* (16:0) in plants and animals, or *stearic acid* (18:0) in animals. These two specific acids are the most common biological form and together comprise approximately 50% of all naturally occurring fatty acids. One of the most common plant membrane phospholipids – phosphatidylcholine (previously termed lecithin) contains palmitic acid at the *sn*-1 site.

As pointed out in further detail in the next chapter this is not an ironclad rule and in certain instances certain amounts of unsaturated fatty acids may also be found at this site.

2.5.b. *At the* sn-*2 site*

In plant membranes *linoleic acid* (18:2) and to a lesser extent *linolenic acid* (18:3) or oleic acid (18:1) are the main fatty acids found at the *sn*-2 site in phospholipids. The very abundant α-linolenic acid – 18:3 $\Delta^{9,12,15}$ accumulates mainly in the galactolipids rather than in the phospholipids. In this respect most plants differ from animals whose

Table 2.2. Nomenclature, sites and type of binding of the major long chained fatty acids occurring in nature (after Cook, 1985)

Biological name	Chemical name	*Abbreviation	Bond positions
Palmitic acid	Hexadecanoic acid	16:0	
Palmitoleic acid	9-hexadecanoic acid	16:1 (*n*-7)	Δ^{9}
	6-hexadecanoic acid	16:1 (*n*-10)	D^{6}
Stearic acid	Octadecanoic acid	18:0	
Oleic acid	9-octadecanoic acid	18:1 (*n*-9)	Δ^{9}
Vaccenic acid	11-octadecaenoic acid	18:1 (*n*-7)	Δ^{11}
Petroselenic acid	6-octadecaenoic acid	18.1 (*n*-12)	Δ^{6}
Elaidic acid	*t*-9-octadecaenoic acid	*t*-18:1 (*n*-9)	*t*-Δ^{9}
Linoleic acid	9,12-octadecadienoic acid	18:2 (*n*-6)	$\Delta^{9,12}$
Linoelaidic acid	*t,t*-9,12-octadecadienoic acid *t,t*-18:2 (*n*-6) *t,t*-$\Delta^{9.12}$		
α-Linolenic acid	9,12,15-octadecadienoic acid	18:3 (*n*-3)	$\Delta^{9,12,15}$
γ-Linolenic acid	6,9,12-octadecadienoic acid	18:3 (*n*-6)	$\Delta^{6,9,12}$
Arachidic acid	Eicosanoic acid	20:0	
Gadoleic acid	9-eicosaenoic acid	20:1 (*n*-11)	Δ^{9}
Gondoic acid	11-eicosaenoic acid	20:1 (*n*-9)	Δ^{11}
Dihomo-γ-linolenic acid	8,11,14-eicosatrienoic acid	20:3 (*n*-6)	$\Delta^{8,11,14}$
Mead acid	5,8,11-eicosatrienoic acid	20:3 (*n*-9)	$\Delta^{5,8,11}$
Arachidonic acid	5,8,11,14-eicosatetraenoic acid	20:4 (*n*-6)	$\Delta^{5,8,11,14}$
Timnodonic acid	5,8,11,14,17-eicosapentaenoic acid	20:5 (*n*-3)	$\Delta^{5,8,11,14,17}$
Behenic acid	Docosanoic acid	22:0	
Centoleic acid	11-docosaenoic acid	22:1 (*n*-11)	Δ^{11}
Erucic acid	13-docosaenoic acid	22.1 (*n*-9)	Δ^{13}
Adrenic acid	7,10,13,16-docosatetraenoic acid	22:4 (*n*-6)	$\Delta^{7,10,13,16}$
Docosapentaenoic acid	4,7,10,13,16-docosapentaenoic acid	22:5 (*n*-6)	$\Delta^{4,7,10,13,16}$
Clupanodonic acid	7,10,13,16,19-docosapentaenoic acid	22:5 (*n*-3)	$\Delta^{7,10,13,16,19}$
Cervonic acid	4,7,10,13,16,19-docosahexaenoic acid	22:6 (*n*-3)	$\Delta^{4,7,10,13,16,19}$
Lignoceric acid	Tetracosanoic acid	24:0	
Nervonic acid	15-tetracosaenoic acid	24:1 (*n*-9)	Δ^{15}

* Unless indicated with a *t* (*-trans*), all bonds have a *cis* formation.

membrane phospholipids principally contain *arachidonic acid* (20:4) at this site. In keeping with the above, Fast Atom Bombardment Mass Spectometry revealed that the *only* phosphatidylinositol (PI) in carrot cells was palmitoyl, linoleoyl-PI whereas animal lipids contained stearic (18:0) and arachidonic (20:4) acids (Van Beemen et al., 1990). Mosquitos have an absolute dependence upon mammals for their requirement of arachidonic acid – mosquito larvae produced by females deprived of mammalian prey, fail to develop. Fig. 2.5 shows

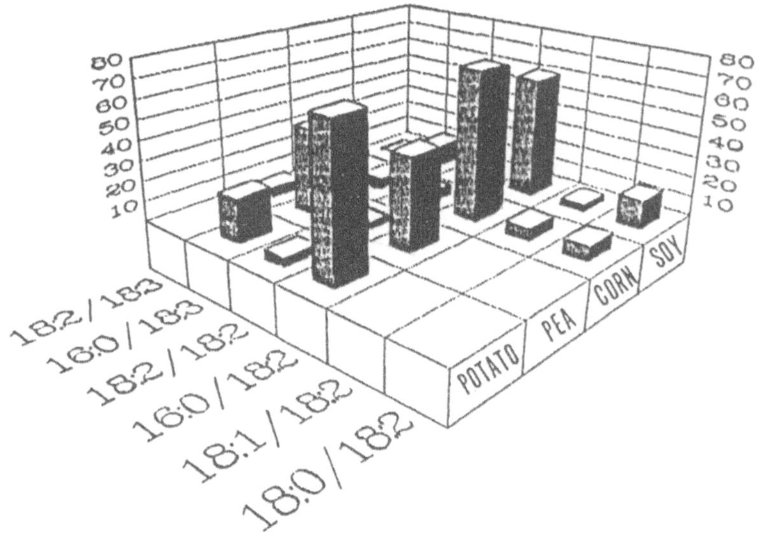

Fig. 2.5. Pattern of PI glycerolipid fatty acyl distribution (percentage) in membranes of various plant species. In the fatty acid column, the first type listed represents the acid present at the *sn*-1, and the second, at the *sn*-2 site. This diagram was kindly provided by Prof. P. Mazliak, Dept. of Plant Physiology, University of Paris.

typical *sn*-1 16:0 and *sn*-2 18:2 prevalence in membrane glycerolipid acyl tails of several plant species.

Mammalian systems cannot introduce double bonds beyond the Δ^9 position: despite this "disadvantage" we know that long chained polyunsaturated acids do occur in animals e.g. the above mentioned arachidonic acid which fully dcscribed is 20:4 $\Delta^{5,8,11,14}$, and others. It therefore is apparent that mammals have an obligatory dietary requirement for fatty acids of plant origin either as first order feeders, or as secondary or tertiary feeders upon other organisms which are themselves first order plant consumers. An example of higher order feeders are carnivorous animals of prey which depend on grazing wild animals for source of food, or of predatory fish which depend on smaller fish that directly feed on phytoplankton. This interdependence is illustrated in Fig. 2.6.

The *desaturase* group of enzymes mentioned above play an important role in assembly of membrane-contained fatty acyl residues. Plants generally introduce the second and third double bonds between the existing double bond and the terminal methyl group. On the other hand, animals introduce the second and subsequent double bonds between an existing bond and the carboxyl end of the acyl chain and never on the methyl end of an existing bond. Thus plants apparently

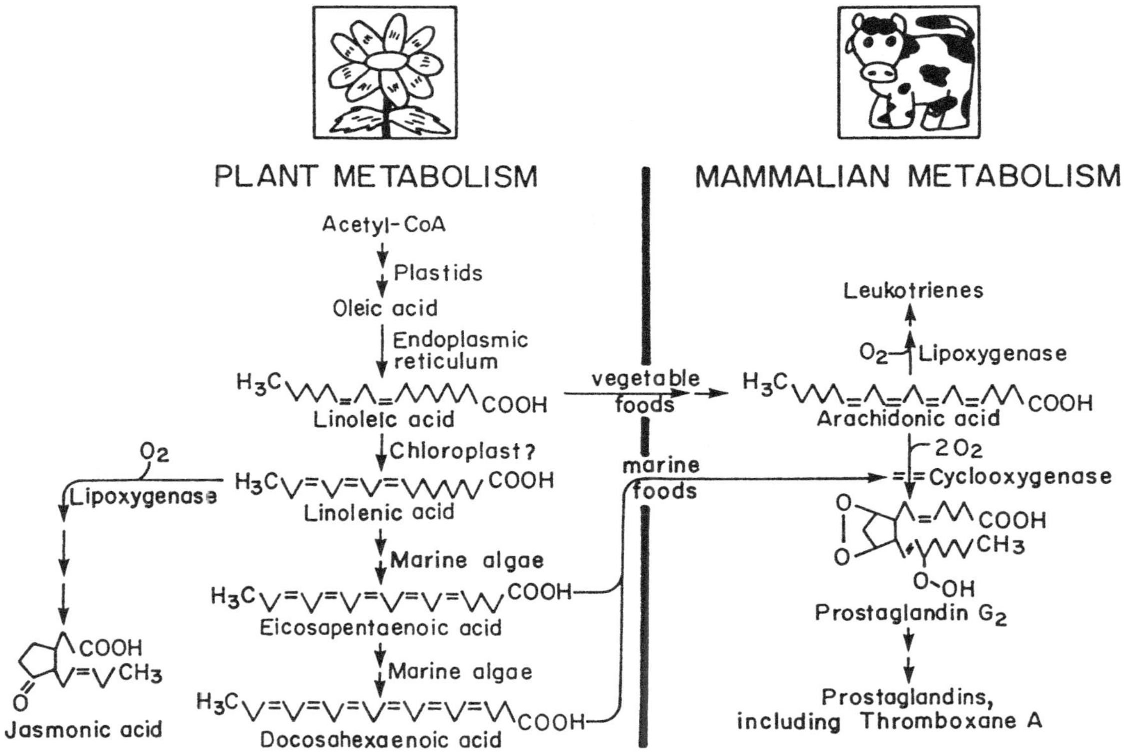

Fig. 2.6. Interdependence and cellular sites of synthesis of polyunsaturated fatty acid in plants and animals (After Kelly, 1984). For simplification the fatty acids are drawn as straight chains without the bends introduced by the *cis* C═C bonds. Further details of some of these pathways are given in Fig. 12.2.

possess Δ^9, Δ^{12} and Δ^{15} desaturases and animals – Δ^9, Δ^6, Δ^5 and Δ^4 desaturases.

Based upon elucidation and verification of some of the above pathways in soybean and *Arabidopsis thaliana* chloroplast membranes, a herbicide BASF 13-338 (San 9785) belonging to the group of substituted pyridazinones has been developed. Norman et al. (1991) working at Beltsville have confirmed that indeed linoleic acid (18.2) and linolenic acid (18:3) are synthesized by a desaturase mechanism by sequential desaturation of 18:1 already esterified to specific glycerolipids. On the one hand polyunsaturated fatty acid formation in the chloroplasts was found by them to be inhibited by the above herbicide, while on the other hand, 18:2 desaturase activity was stabilized by antioxidants.

2.6. *n*-3 and *n*-6 fatty acids in plant disease

A special case of effect of long chained fatty acids of the *n*-3 family – EPA, DPA and DHA (see page 20) and of arachidonic acid of the *n*-6 group is the eliciting of resistance to potato blight (*Phytophtera*) stemming from the production of phytoalexins. The research of Kuç's group at the University of Kentucky has shown that phytoalexin production is a response reaction of host tissue (not necessarily membrane-located) to the presence of the above *n*-3 and *n*-6 fatty acids in the mycelia of the invading *Phytophtera* (Zook and Kuç, 1987). Ozeretskovskaya et al. (1988) and also Merzliac of Moscow University have verified these findings and have added a further detail: the above reactions take place upon interaction of these fatty acids with a *lipoglycoprotein* situated in the pathogen. These workers have moreover indicated that such acquired resistance is found in tomatoes and sugar beet as well as in potatoes, and involves a certain amount of immunity not only to *Phytophtera* blight but also to other fungal and bacterial pathogens. Cohen et al. (1991) have found that spray-applied long chained unsaturated fatty acids may impart systemic resistance to the potato blight fungus, the order of effectivity being 20:5 > 20:4 > 18:2 > 18:3 > 18:1. The mechanism involved in the observed systemic protection of potato plants is as yet unknown.

Regarding the problem as related to fatty acid metabolism from the opposite aspect, i.e. infection mechanisms and not immunity, there is much current interest in *jasmonic acid* and its metabolites such as *methyl-jasmonate*. As seen in Figs 2.6 and 12.2 and discussed in greater detail in Chapter 12 this is a catabolite of the *n*-3 linolenic acid (18:3) and, as discussed earlier in this chapter, to a certain extent may be regarded as the plant analogy of mammalian prostaglandins (Heinberg et al., 1988; Parthier, 1990; Sembdner et al., 1990), jasmonic acid also being closely associated with ageing processes (Ueda and Kato, 1980) and especially with Ca^{2+}-triggered membrane senescence (Leshem, 1987).

2.7. References and recommended reading

Brenner, R.R. 1984. Effect of unsaturated fatty acids on membrane structure and enzyme kinetics. *Prog. Lip. Res.* 23:69–96.

Caroll, K.K. 1986. Biological effects of fish oils in relation to chronic diseases. *Lipids* 21:731-32.

Cohen, Y., Gisi, U. and Mosinger, E. 1991. Systemic resistance of potato plants against *Phytophtera infestans* induced by fatty acids. *Physiol. Mol. Pl.. Path.* 38:255-63.

Cook, H.W. 1985. Fatty acid desaturation and chain elongation. pp. 181–212. In: Eds. D.E. Vance and J.E. Vance. *Biochemistry of Lipids and Membranes.* Benjamin Cummings Pub. Co., California.

de Jong, T. 1980. *Triacylglycerol crystal structures and fatty acid conformations.* University of Utrecht, Holland. pp. 1-0171 Ph.D. Thesis.

Evans, D.F. 1988. Self organization of amphiphiles. *Langmuir* 4:3–12.

Gurr, M.I. and James, A.T. 1980. *Lipid Biochemistry: An Introduction.* Chapman and Hall, London, New York.

Harwood, J.L. 1987. Lung surfactant. *Prog. Lip. Res.* 26:211–56.

Heinberg, M., Miersch, O. and Sembdner, G. 1988. Absolute configuration of 12-oxo-10, 15(2) phytodienoic acid. *Lipids* 23:521–4.

Kelly, G.J. 1984. Formation and fates of plant poly-unsaturated fatty acids. *TIBS*, December edition, 501–3.

Leshem, Y.Y. 1987. Membrane phospholipid catabolism and calcium activity in the control of senescence. *Physiol. Plant.* 69:551–9.

Leshem, Y.Y., Cojacaru, M. and Landau, E.M. 1990. Biophysical compaction studies of effects of abscisic acid on membrane phospholipids: possible role in stomatal physiology and senescence. *Biochem. Physiol. Pflanz.* 186: 351–55.

Leshem, Y.Y., Landau, E.M. and Deutsch, M. 1988. A monolayer model study of surface tension-associated parameters of membrane phospholipids – effect of unsaturation of fatty acyl chains. *Jour. Exp. Bot.* 39:1679–87.

Norman, H., Pillai, B. and St. John, J. 1991. Herbicide interactions in the biochemical and chemical regulation of plant polyunsaturated fatty acids. pp. 399–406. In: Eds. P.J. Quinn and J.L. Harwood. *Plant Lipid Biochemistry, Structure, Utilization and Function.* Portland Press, London and Colchester.

Ozeretskovskaya, O., Chaleva, L., Avdyushko, S., Chalenko, G. and Karavaeva, I. 1988. Eicosanoids as inducers of enhanced resistance of potato to late blight. *Soviet Plant Phys.* 35:141–48.

Parthier, B. 1990. Jasmonates: hormonal regulators or stress factors in plants? *Plant. Gr. Reg.* 9:57–63.

Sanders, T. 1988. Essential and *trans* fatty acids in nutrition. *Nut. Res. Rev.* 1:57–78.

Sanders, T. 1990. The pros and cons: polyunsaturated and monounsaturated fats. *Chem. Indust.* 13:427–30.

Sembdner, G., Meyer, A., Miersch, O. and Brückner, C. 1990. Metabolism of jasmonic acid. pp. 374–9. In: Eds. R.P. Pharis and S.B. Rood. *Plant Growth Substances 1988.* Springer Verlag, Berlin, Heidelberg.

Stymne, S., Bafor, M., Jonsson, L., Wiberg, E. and Stobart, K. 1991. Triacylglycerol assembly. In: Eds. P.J. Quinn and J.L. Harwood. *Plant Lipid Biochemistry, Structure, Utilization and Function.* Portland Press, London and Colchester (In press).

Tarchevsky, I.A., and Grechkin, A.N. 1989. Perspectives of search for eicosanoid analogs in plants. pp. 45–9. In: Eds. P.A. Biacs, K. Gruiz and T. Kremmer. *Biological Role of Plant Lipids.* Plenum Press, New York and Akademiai Kiado, Budapest.

Tasaka, Y., Nishida, I., Higashi, S., Beppu, T. and Murata, N. 1990. Fatty acid composition of phosphatidylglycerols in relation to chilling sensitivity of woody plants. *Plant. Cell. Physiol.* 31:545–50.

Trémolières, A., Dubacq, J.P., Duval, J.C., Lemoine, Y., and Remy, R. 1982. Role of PG-containing *trans*- 16:1 acid in oligomeric organization of the light harvesting chlorophyll protein. pp. 369–73. In: Eds. J. Wintermans and P. Kuiper. *Biochemistry and Metabolism of Plant Lipids.* Elsevier, Amsterdam.

Ueda, J. and Kato, J. 1980. Isolation and identification of a senescence promoting substance from wormwood (*Artemisia absinthium* L.). *Plant Phys.* 66: 246–249.

van Beemen, R.B., Wheeler, J.J. and Boss, W. 1990. Identification of carrot inositol phospholipids by fast atom bombardment mass spectrometry. *Lipids* 25:328–34.

Wada, H., Gombos, Z. and Murata, N. 1990. Enhancement of chilling tolerance of a cyano-bacterium by genetic manipulation of fatty acid desaturation. *Nature* 347:200–3.

Zook, M. and Kuç, J. 1987. Arachidonic and eicosapentaenoic acids, glucans and calcium as regulators of resistance to plant disease. pp. 75–82. In: Eds. P.K. Stumpf, J.B. Mudd and W.D. Ness. *The Metabolism, Structure and Function of Plant Lipids.* Plenum Press, New York.

3. Membrane glycerolipid and sphingolipid headgroups

The glycerolipid headgroups serve as the hydrophilic plane of the membrane bilayer and on the one hand provide the interface between the membrane and its milieu while on the other hand, between the internal cytosol and the membrane. In the previous chapter we pointed out that the "backbone" of glycerolipids is the alcohol – glycerol. In the various *phospholipids* the third serial number carbon atom (*sn* – 3) undergoes esterification with a phosphate (PO_4) group to initially form *phosphatidic acid*. The latter is considered the parent compound from which other species of phospholipids may be derived. The *sn* – 3 carbon atom may also bind with either a carbohydrate group (primarily galactose or glucose) or with a sulphur containing constituent in which case respectively *galactolipids* or *sulpholipids* are formed. In sphingolipids the basic alcohol is not glycerol but rather *sphingosine* and the anchoring function joining the headgroup to the two non-polar tails is fulfilled by the serine-derived section forming the initial two carbon atoms of this long chained alcohol.

Following is a more or less systematic classification and description – inclusive of physiological properties – of the major glycero- and sphingolipid species found in higher plant membranes. This subject is dealt with under the following headings:

1. *Electroneutral Phospholipids* – phosphatidylcholine, phosphatidylethanol, phosphatidylethanolamine.
2. The *Anionic Phospholipids* – phosphatidylserine, phosphatidylglycerol, the phosphatidylinositides.
3. The *lyso-Phospholipids*.
4. *Phospholipid Asymmetry and Symmetry in Membranes* – biochemical asymmetry, electrochemical asymmetry.
5. The *Sphingolipids* – cerebrosides, sphingomyelins and gangliosides, biophysical effects.
6. *Chloroplast-specific Glycerolipids* – galactolipids, sulpholipids.
7. *Mitochondrial Phospholipids* – including cardiolipin.
8. *Platelet Activating Factor*.
9. The *Tertiary Structure of Membrane Glycerolipids*.

3.1. Electroneutral phospholipids

3.1.a. *Phosphatidylcholine (PC – also termed lecithin)*

Barring the case of chloroplasts, PC is the most abundant phospholipid component in both animal and plant membranes. Besides glycerol which is one type of alcohol, PC also contains another alcohol – *choline*. The upshot therefore is that PC contain two alcohol groups. Concerning its electric charge, (which as will be later discussed in detail is of primary biophysical importance in membrane function) it has a net neutral charge – PC being a classic example of a *Zwitterion* (*zwitter* in German meaning hybrid). This is because the positive charge on the choline is neutralized by the negative charge of the phosphate (See Fig. 3.1 and Table 3.3 at the end of this chapter). It is pointed out that PC is not always uniform in composition since while the choline headgroup is always present, a wide heterogeneity may exist in the nature of its two fatty acyl chains.

Because of its abundance and amphiphilic properties, the hydrophilic nature of its headgroup and the lipophylic nature of the fatty acyl chains, PC, usually under its popular name *lecithin* is widely used as an emulsifier in the food industry. Erdélyl (1989) has pointed out that *lyso*-PC, i.e. PC treated with phospholipase A_2 is even a better emulsifier than the intact PC and that more than half of commercial soybean lecithin principally employed in the Western European food industry is pretreated with porcine phospholipase A_2.

3.1.b. *Phosphatidylethanol (PE)*

The PE molecule in its entirety also contains two alcohol groups – the primary glycerol as well as the secondary ethanol of the headgroup. Here again the positive headgroup charge balances out with the negative phosphate charge of the "neck" section to produce a net electroneutral Zwitterionic phospholipid.

3.1.c. *Phosphatidylethanolamine (PEA)*

Like PC and PE, PEA also contains two alcohols. The amine is conjugated to the ethanol in the headgroup. A glance at the structure of PEA (Fig. 3.1) reveals that in terms of electrochemistry a difference exists bctween PEA and PC even though both are neutral species.

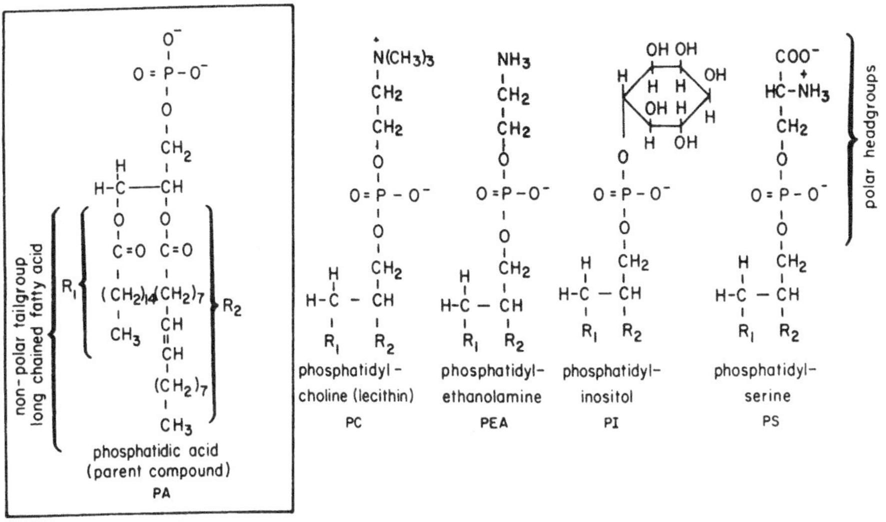

Fig. 3.1. The main phospholipid categories of the plant membrane. In the headgroup of these species a second alcohol which can be choline, ethanolamine or inositol can be present. In phosphatidylserine, the amino acid serine replaces the secondary alcohol. R_1 and R_2 are fatty acids esterified to the hydroxyl groups of the glycerol (* PI derivatives are detailed in Fig. 3.2).

The difference lies in the fact that the positive charge of the amine in PEA is primary, whereas in PC the positive charge is secondary, not being situated at the "top" of the headgroup but somewhat internalized and surrounded by three methyl (CH_3) groups. What transpires from this is that PEA in the membrane is concomitantly more amenable to crosslinking or electrostatic binding than PC and resultingly imposes a greater rigidity on membranes than PC.

In many membrane species, especially in mammals the above electroneutral phospholipids are believed to be more concentrated on the outer monolayer leaflet of the biological bilayer. The moot topic of symmetric or asymmetric phospholipid distribution in plant and mammalian membranes is dealt with later under section 3.4 in the present chapter.

3.2. The anionic phospholipids

3.2.a. *Phosphatidylserine* (*PS*)

The headgroup of this phospholipid species contains the hydrophilic amino acid *serine* whose negative charge, in addition to the negative charge on the phosphate neck section, causes PS to be pronouncedly anionic. In the erythrocyte membrane and possibly others as well, PS is located on the inner monolayer of the bilayer and together with other anionic-PL species is believed to contribute to the "trans-membrane potential" (cf. section 3.4).

PS is moreover the predominant acidic phospholipid in a number of biological membranes, is involved in physiological processes such as membrane fusion, regulation of homeostasis and the activation of protein kinase C. In these processes PS acts in concert with Ca^{2+} for which it has high affinity. The asymmetric distribution of PS over the inner and outer leaflet of cell membrane has been suggested to be functional in the interplay of the membrane with extrinsic membrane proteins and cytoskeletal elements. Moreover, PS is an important model phospholipid for the study of lipid-protein interactions in which a major role has been attributed to anionic (acidic) phospholipids (de Kroon et al., 1990).

3.2.b. *Phosphatidylglycerol* (*PG*)

In chloroplast thylakoid membranes very little phospholipid is found, but the small amounts of what is detected nevertheless is primarily PG which comprises ca 10% of their total glycerolipid, and approximately the same content of this anionic species is present in mitochondrial membranes. In PG the second alcohol group is glycerol. Interestingly the fatty acid tails also contain rare species of long chained unsaturated acid, these being a 16:3 and a *trans* 16:1 Δ^3 species. A particular type of a chloroplast membrane-contained PG – dipalmitoyl PG – or a *sn*-1 palmitoyl, *sn*-2 *trans* 16:1 PG has been found by Murata's group (cf. Tasaka et al., 1990 and references therein) to be associated with *cold acclimation* of herbaceous plants as well as of deciduous and evergreen tree species. A like trend has been found by Browse in *Arabidopsis* when grown under low light intensities. The physiological and biophysical implications of these findings which may reflect an opposite trend to that usually followed whereby cold acclimation involves increased desaturation (discussed in detail in Chapter 10) are at present not known.

3.2.c. *The phosphatidylinositides*

This phospholipid group includes phosphatidylinositol (PI), phosphatidylinositolphosphate (PIP_2), and phosphatidylinositol-

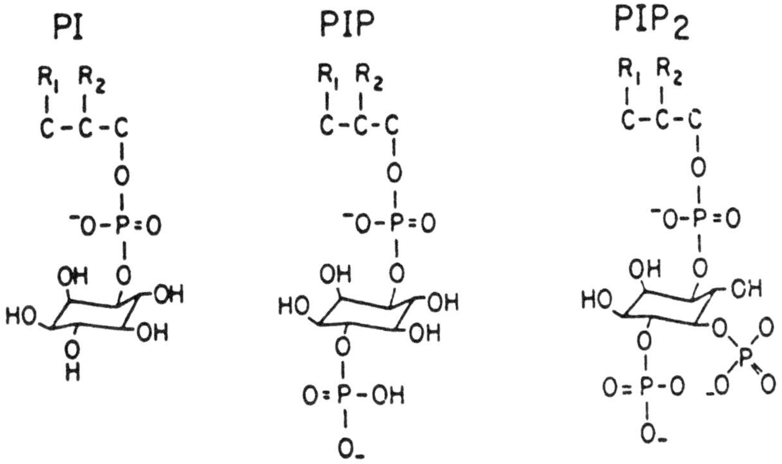

Fig. 3.2. The phosphatidylinositides. The group comprises only 10–15% of total membrane lipids but because of their markedly anionic character and the increasing number of negative charges with increase of phosphate groups, they fulfill an important role in several pivotal regulatory processes including electrostatic cross-linking with divalent cations like Ca^{2+} and Mg^{+2}, signal transduction, protein kinase activation, and when appearing on the inner leaflet, contribute to the negative transmembrane potential.

trisphosphate (PIP_3). See Fig. 3.2. As a group the phosphatidylinositides have a cyclic sugar alcohol – *inositol* – or phosphorylated derivatives of inositol conjugated to the phosphate neck section. With increase of phosphorylation and subject to pH effects (van Paridon et al., 1986; Landau and Leshem, 1988) these phospholipids become increasingly negatively charged. Chapter 12 of this book will deal with the pivotal biological importance of this group of anionic phospholipids, which while being a minor membrane component plays a major regulatory role. PI is also an important structural component of the inner mitochondrial membrane (see Fig. 3.11). It is of interest that a legume seed-reserve component – *phytic acid** – is no other than *inositol-hexaphosphate* in which all six of the hydroxyl groups of the inositol are replaced by phosphate groups.

In mammalian plasma membranes phosphatidylinositol has been documented to be associated with the inner monolayer of the bilayer, however exceptions do occur, a notable one being neural synapses of the electric eel (*Torpedo* spp) where PI is the *sole* component of

* *Phytic acid* may comprise 0.75–1.25% of total dry weight of legume seeds. The well known culinary problem of "hard-to-cook" starchy legumes is believed to be associated with phenol-mediated binding of phytic acid to cell wall components (Batty and Slinkard, 1989). However as yet no evidence has been produced that this hexaphosphate inositol is located in the membranes of these species.

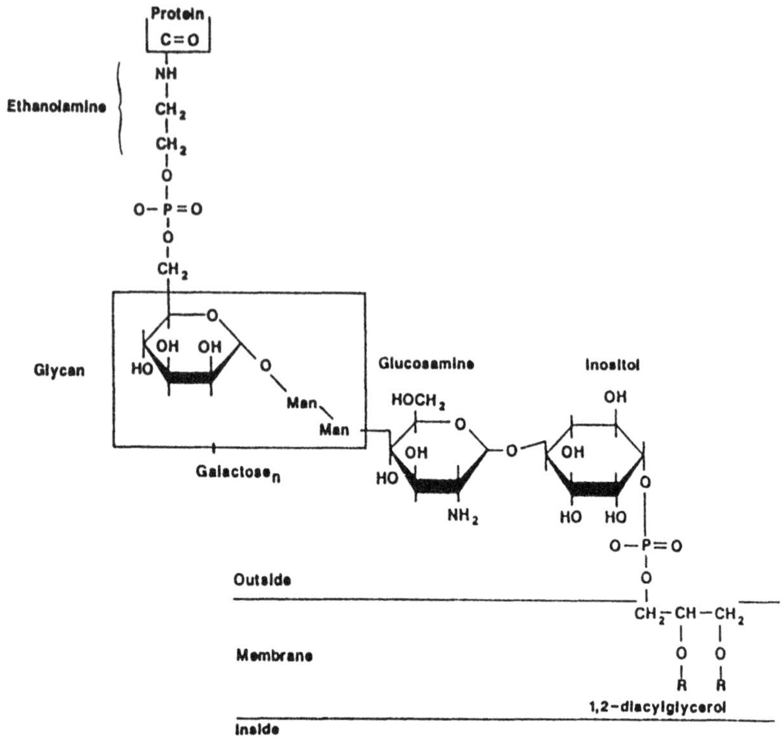

Fig. 3.3. The structure of the glycosyl-phosphatidyl anchor between the membrane and extramembranal carbohydrates and proteins (After Low, 1987).

the outer membrane leaflet. Another important function of PI, well documented in mammalian membranes (Low, 1987; Olson, 1988), but to best of our knowledge not yet reported in plant membranes, is the *anchoring*, via a glycerol linkage, of sections of membrane proteins which extend beyond the bilayer (see Fig. 3.3).

3.3. The *lyso*-phospholipids

Lyso-phospholipids may be defined as phospholipids from which, due to phospholipase enzyme action, one of the two fatty acyl has been removed (Fig. 3.4). Since membranes are dynamic systems and undergo micro- or macro-environmental changes including degree of saturation, or length of fatty acyl side chains, and are subject to phospholytic breakdown, it can be assumed that at any given developmental stage at least some of the aforementioned phospholipid species may exist in the form of a *lyso*-phospholipid from which one of the side chains has been removed.

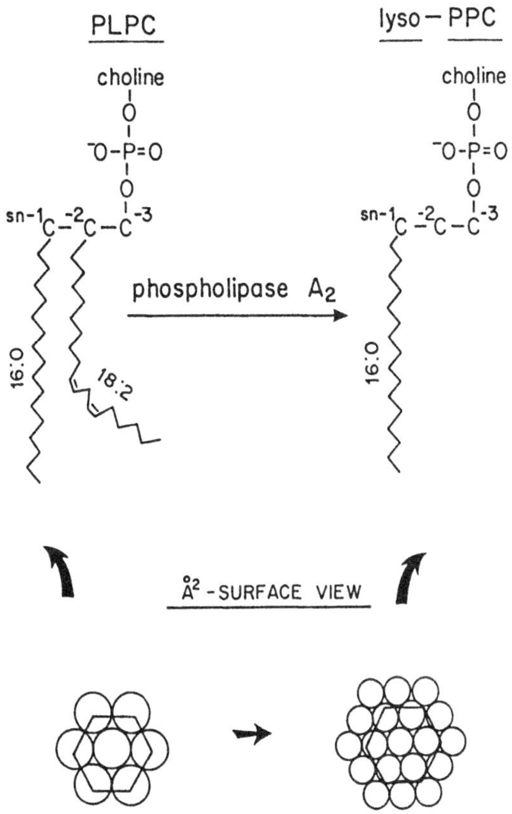

Fig. 3.4. A typical lyso-phospholipid – palmitoylphosphatidyl choline (PPC) obtained after deesterification by phospholipase A_2 of linoleoyl-palmitoyl-PC (LPPC). *Lyso*-phospholipids play a role in "retailoring" of membranes in response to environmental change and stress.

In mitochondrial membranes Beatrice et al. (1980) have pointed out that the underlying mechanism involved in catabolic permeability increase, collapse of membrane potential and progressive acceleration of respiration is *lyso*-phospholipid rather than free polyunsaturated fatty acid although the latter may have a synergistic effect. Moreover in terms of biophysical configurations de Gier et al. (1982) indicate that *lyso*-phospholipids possessing only one fatty acyl tail can result in physiologically detrimental formation of the non-bilayer hexagonal I (H_I) configuration (see Chapter 7). On the other hand these *lyso*-phospholipids, if still located in the bilayer, may serve as cornerstones for remodelling.

In adaptation of plants to stress, especially related to temperature, it is currently believed that one of the regulatory mechanisms involves change in phospholipid constituency, not by synthesis of new phospho-

Table 3.1. Surface tension values of phospholipid and *lyso*-phospholipid monolayers (from Leshem, 1991)

Monolayer	Surface tension (mNm^{-1})
sn-1 palmitoyl, *sn*-2 linoleoyl-PC (PLPC)	30.52
lyso-sn-1 palmitoyl-PC	38.63
PLPC+phospholipase A_2*	36.35

*Deesterifies the *sn*-2 fatty acyl chain

lipid molecules but rather by "retailoring" of existing ones. This may be achieved by phospholipase-mediated release of either the *sn*-1 or *sn*-2 fatty acyl chain and reinsertion of a fatty acid residue better equipped to withstand environmental change – increase of saturation at higher temperatures and decrease of saturation to withstand low temperature – in both cases the trend being a homeostatic mechanism to sustain optimal membrane fluidity.

Lyso-phospholipids may also act as surfactants, disrupt bilayers at concentrations far below those required for membrane dissolution, aggregate into micelles at much higher Critical Micelle Concentrations than parent phospholipids (Stafford and Dennis, 1988). They are moreover involved with membrane fusion and being cone-shaped are believed to dominate in membrane surfaces of high curvature, these possibly including the highly folded regions of thylakoid membranes and ER (Lucy, 1970). *Lyso*-phospholipids, especially *lyso*-PC also affect membrane contained proteins such as Na^+/K^+ ATPase (Weltzien, 1979) and plant plasmalemma-contained H^+ ATPase (Palmgren et al., 1988). These effects may be caused by partitioning substrate out of bulk phase into a phase of different accessibility to the enzyme, or by inducing membrane surface tension changes which allow enzyme mobility. This latter biophysical principle is illustrated in Table 3.1.

3.4. Phospholipid asymmetry and symmetry in membranes

This topic will be discussed under two headings:

1. Biochemical asymmetry and symmetry.
2. Electrochemical asymmetry.

The latter subsection extends beyond the realm of phospholipids but for methodological reasons we considered the present location as appropriate.

3.4.1. *Biochemical asymmetry and symmetry*

It would be tempting to provide the reader with a picture of plant membranes that resembles those of mammalian ones. Before presenting the very partial knowledge on the plant membrane situation, that in the mammalian membrane is briefly described.

The two individual leaflets comprising mammalian membranes exhibit *absolute* asymmetry of proteins and carbohydrate moieties of glycoproteins, but only *partial* asymmetry of lipids. In other words, a given phospholipid is normally found in both phospholipid monolayers, but at unequal concentrations. Lipid asymmetry has been detected by exposing membranes to membrane-impermeant compounds reacting with the-NH_2 groups of PE, to phospholipases, or to phospholipid exchange proteins. Different membranes differ dramatically in the asymmetric arrangement of their phospholipids (Schroeder, 1984). For example, 63 to 75% of PC is localized in the outer monolayer of the erythrocyte membrane whereas most of PE and all of the PS are in the inner (i.e. cytoplasmic) monolayer (Fig. 3.5). In erythrocytes this lipid asymmetry causes the outer leaflet of the plasma membrane to be less "fluid" than the inner one (Seigneuret et al., 1984). Similarly, two thirds of the PE found in the (single) cytoplasmic membrane of the gram-positive *Bacillus megaterium* are in the inner monolayer. A situation resembling the asymmetry in the human erythrocyte is also found in the influenza virus. How this asymmetry is maintained is not clear. Evidence from experiments with erythrocytes suggests that PE-PS asymmetry is maintained by an ATP-dependent mechanism or that the cytoskeleton is involved (Geiger, 1985). Slow "*flip-flop*"* of phospholipid cannot be the sole cause of the asymmetry as increases in the "flip-flop" rate do not alter asymmetry. A transmembrane electrical potential is possibly involved but this remains to be proven.

The final outcome of all of the above is that the inner (or endofacial) leaflet of the mammalian plasma-membrane is of a more anionic

* Besides describing ear movements of the Cocker Spaniel and similar dog breeds, "*flip-flop*" is a term used to describe the process whereby an intact phospholipid may traverse the bilayer and shift from one monolayer leaflet and relocate in the other. "Flip-flop" is believed to be aided by an enzyme designated *flippase*, evidence for whose existence was presented by Bishop and Bell (1985). In metabolically inactive membranes such as liposomes and virus membranes "flip-flop" activity is markedly low, whereas in active membranes activity is high and is even more pronounced in cancer cells. Y.Y.L.

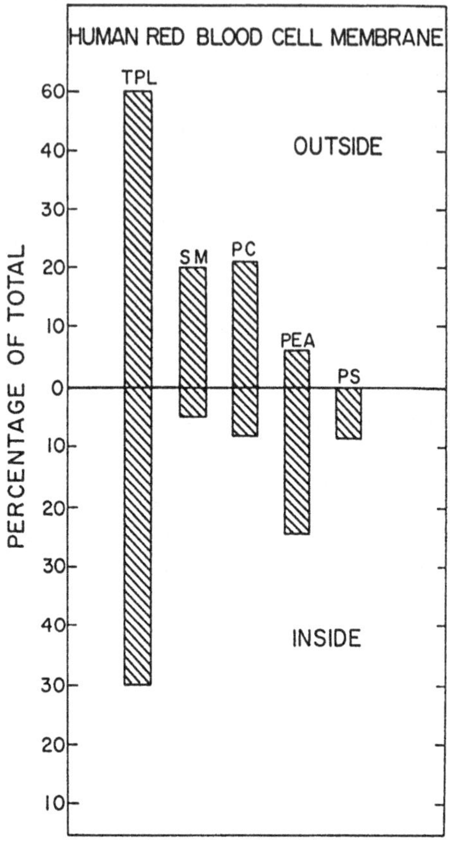

Fig. 3.5. Asymmetrical distribution of phospholipids in membranes of human red blood cells. TPL = total phospholipid. SM = sphingomyelin.

nature than the outer (or extrafacial) one and this phenomenon constitutes the electrostatic +/ − gradient discussed in a later section.

As opposed to the situation in mammalian membranes, very little is known about the degree of asymmetry in plant membranes, and most evidence suggesting that there might be some analogy is mainly indirect and stems from differential responses of plant membranes to exogenously applied chemicals and reagents. Evidence to this effect may be obtained from information gathered from the different effects of the divalent Ca^{2+} ion on the plasma membrane of stomatal guard cells. As shown by McAinsh et al. (1990) when acting in contact with what is assumedly the inner leaflet of the bilayer, Ca^{2+} triggers a classic *signal transduction* effect leading to stomatal closure. The same divalent Ca^{2+} ion acting on the guard cell membrane exterior, i.e. when Ca^{2+} is *exuded* from the apoplast and comes in contact with the guard cell membrane outer leaflet, inhibits stomatal closure in response to abscisic acid (Mansfield and Atkinson, 1989).

Likewise, in senescence both of leaves and fruit, Leshem et al. (1986) have reported that Ca^{2+}, when located in the free space and being in contact with the outer monolayer of the bilayer has a marked senescence *retarding* effect, whereas Ca^{2+} after having crossed the membrane and entered the cytosol has a pronounced senescence *promoting* effect associated with catabolic processes occurring in the inner leaflet. In this respect the phenomenon somewhat resembles the events described in the stomatal guard cell as discussed above. Phospholipid asymmetry and human ageing especially related to neurophysiology is thoroughly reviewed by Schroeder (1984).

Notwithstanding the paucity of plant reports there nevertheless is some direct evidence for plant membrane glycerolipid asymmetry: some concise work in this respect has been presented by Siegenthaler's group at Neuchâtel (cf. Rawyler et al., 1990) where it has been elegantly demonstrated that thylakoid membranes are asymmetric and that this pertains to monogalactosyldiacylglycerol (MGDG) export from the chloroplast envelope membrane to the thylakoid membrane and also to the biogenesis and maintenance of leaflet MGDG asymmmetry in the thylakoids.

More work is still necessary to determine whether or not plant membranes follow the classic mammalian asymmetry pattern. An often-heard statement is that since degree of plant differentiation is markedly less than that of animals, so is membrane-phospholipid asymmetry less. This assertion should be taken into account but treated with some degree of reserve at present.

3.4.2. *Electrochemical asymmetry*

A far clearer picture of electrochemical asymmetry in plant membranes exists and here it is becoming increasingly evident that in many plant membrane species, as in mammalian ones, the inner section of the membrane is negative relative to the outer section: this stemming from a *membrane potential* ($\Delta\psi$) which usually ranges from -10 to -100 mV.

This trans-membrane potential should be distinguished from the membrane *surface potential* which is also negative but far less so than $\Delta\psi$, being in the vicinity of -8 to -30 mV as compared to considerably greater $\Delta\psi$ values of -10 to -100 mV or even more. The membrane surface potential is the electrostatic potential at the membrane aque-

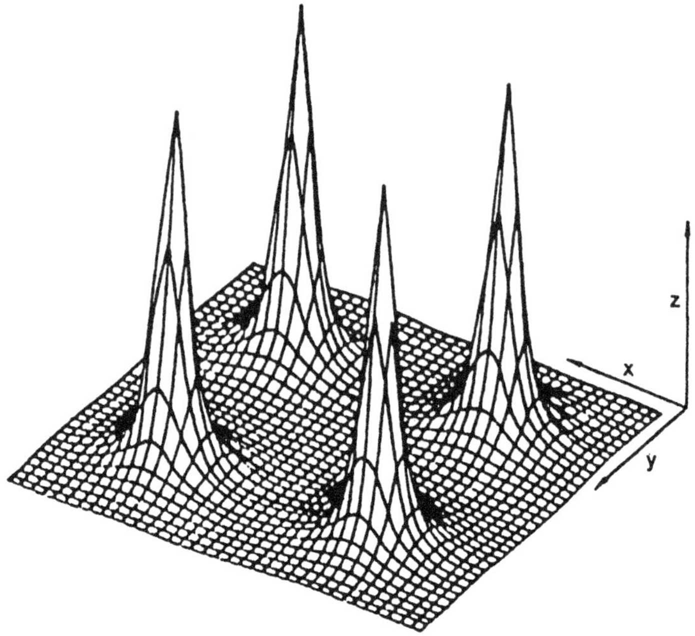

Fig. 3.6. Discreteness-of-charge effect: Negative electrostatic potential profile of a bilayer with discrete surface charge distribution on a square lattice (after Cevc and Marsh, 1987, cf. references to Chapter 1).

ous interface relative to that in the corresponding outer bulk solution and arises from a net charge on the membrane surface. Both negatively charged lipids *and proteins* contribute to the net surface charge density. Honig (1986) has calculated that biological membranes typically have 10–20% of their effective lipid area composed of negatively charged lipids, thus providing about 1 electronic charge per 1000 $Å^2$ membrane surface area. Surface density charges in plant membranes vary with type of organelle and with botanical species (Körner et al., 1985).

Maintenance of a cell in relatively high negative state and not necessarily ceaseless chemical activity is even regarded by some biologists as a basic manifestation of the living state. Functional activities are transient reversible assumptions of a lower (negative) energy state; death, a permanent irreversible one (Ling, 1969).

Thus in terms of membrane topography it is noteworthy that the negative charges cannot be considered to be spread out evenly as a uniform continuum across the monolayer(s) but rather are localized at discrete sites and as such, generate a certain surface-charge density. This is graphically presented in Fig. 3.6. In red blood cells 30% of the inner monolayer is negatively-charged with PS which produces a considerable negative surface potential.

Electro-asymmetry could stem from:

1. The presence, on the plasma of a H^+ATPase membrane which pumps protons from the cytoplasm into the apoplastic space thus creating an electrochemical gradient across the plasma membrane. An interesting case of marked Na^+/H^+ATPase activity is that of the membranes of the salt teargland of bird species such as seagulls which depend upon sea water for drinking. Here a marked ATPase-mediated pumping of salt ions out of the cell is necessary to enable the birds to utilize the otherwise unpotable sea water.

Modulation of plasma bound ATPase depends on factors connected with the supporting lipid bilayer. These factors include firstly the microviscosity of the bilayer's fatty acyl tails – especially their degree of unsaturation – and secondly the putative existence of a lipid *annulus* or a *boundary lipid* which is claimed to encase the enzyme, and so doing anchor it into the bilayer (Cook and Burden, 1990).

The endogenous growth substance indoleacetic acid (auxin) plays a central regulatory role by stimulating a rapid increase in proton secretion by ATPase(s) in the plasma membrane. The resulting proton and voltage gradients across the membrane provide for the transport of ions and other substrates needed in growth and development including amino acids, sugars, and auxin itself (Clark and Goldsmith, 1987). In actively transporting cells such as root hairs of *Sinapis alba* the plasma membrane proton pump consumes as much as 1/4 to 1/2 of the cellular ATP (Felle, 1982). The topic of ATPase and proton pumping in the tonoplast and plasma membranes of stomatal guard cells is discussed in detail in Chapter 11.

2. The nature of the headgroup charge on either side of the membrane, a typical situation of which is presented in Fig. 3.7.

This situation also contributes to the setting up of a *redox system* that transfers electrons from reducing agents in the cytosol to external impermeable oxidants. The redox system of plant plasma membranes has been implicated in several processes including H^+ efflux, K^+ uptake, and iron reduction (Møller and Lin, 1986). More recently Morré et al. (1988) have elegantly shown that plant plasma redox constituents respond to phytohormones of the auxin category and that this hormone response is geared to the membrane-located NADH-oxidase activity. The reader interested in plasma-membrane redox processes in greater detail, including aspects beyond the confines of this text, is referred to the thorough review of Møller and Crane (1990).

In passing we mention that activation of trans-membrane potential has been included in an overall *action potential hypothesis* of plant

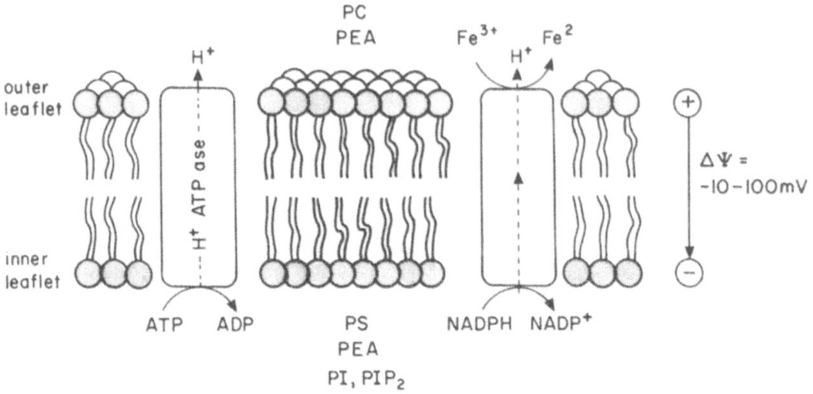

Fig. 3.7. The trans-membrane potential. It is believed that most biological membranes are asymmetric. The outer monolayer leaflet of the bilayer mainly contains PC, PEA and GlCer which are electroneutral, while the inner leaflet is composed primarily of the electronegative phospholipid species – PS, PI and PIP_2. The H^+ATPase protein pump extrudes H^+ ions from the interior as does a NADPH-linked "redox" system. Trans-membrane K^+ and Na^+ gradients also contribute to the gradient. All of the above combine to provide a trans-membrane potential ($\Delta\psi$) varying from minus 10 to minus 100 mV, although under certain circumstances it may be considerably greater. In plants it has been claimed that due to a lesser degree of differentiation as compared to mammals, the asymmetry is less pronounced.

wound response. According to the main protagonist of this hypothesis – Eric Davies of the University of Nebraska – electrochemical action potentials play a major role in intercellular and intermolecular communication and in addition to their role in wounding they may also be important signalling mechanisms for chilling injury, mechanical perturbation, invasion by pathogens as well as for light and gravity (Cf. Davies, 1987). A lucid, if somewhat controversial, exposition of this concept is provided by Davies (1990).

3.5. The sphingolipids

After phospholipids, sphingolipids are the second largest class of membrane lipids. They basically resemble phospholipids in that they possess:

(a) *a polar head* – often D-galactose, glucose, phosphocholine or – ethanolamine – which is hydrophilic.

(b) *two non polar tails* –

(i) the alcohol sphingosine or one of its derivatives (Fig. 3.8)
(ii) a long chained fatty acyl residue.

There are three classes of sphingolipids, the most important of which, and as presently known the sole one detected in plant membranes, being the first. In order to provide a general perspective all three classes are described – the one present in plant species, in greater detail:

1. cerebrosides
2. sphingomyelins
3. gangliosides

3.5.1. *Cerebrosides (also termed glycosphingolipids or glucosylceramides)*

Cerebrosides (Fig. 3.8 above) contain no phosphorous and have no electrical charge since their polar heads are neutral. This class may be included under the category of *galactolipids* since the headgroup is usually *D-galactose* which is of common occurrence in the plant plasmalemma, tonoplast and in brain tissue. *Gluco-cerebrosides* i.e. with a D-*glucose* headgroup are found in non neural tissues. Multilayer (*di-*, *tri-* and *tetra-*) cerebrosides have also been reported e.g. a digalactosyl ceramide has been found in chloroplast membranes. In mammalian tissues these sphingolipids are exclusively located on the outer leaflet of the bilayer: no information on this aspect at present is available for plant membranes.

Although the existence of such sphingolipids in plants was documented more than 30 years ago, comparatively little is known about their structure, function and metabolism in plant cells. However some information on physiological function has been forthcoming – concise research carried out by Liljenberg's group in Sweden (cf. Norberg et al., 1991) has indicated that during the acclimation of oat seedlings to moderate water stress a marked drop of root plasma membrane glucosylceramides occurs and that this is linked to different polymorphic phase behavior of the plasmalemma. Analyses of highly purified plasma membrane and tonoplast preparations by several groups have demonstrated that glucosylceramide (GlcCer), the predominant sphingolipid in plants, is a major component of these two membrane types. This is similar to the situation in animal and fungal cells, suggesting some common basic function for these lipids. Recent studies (Lynch, 1991) aimed at understanding the structure, function and

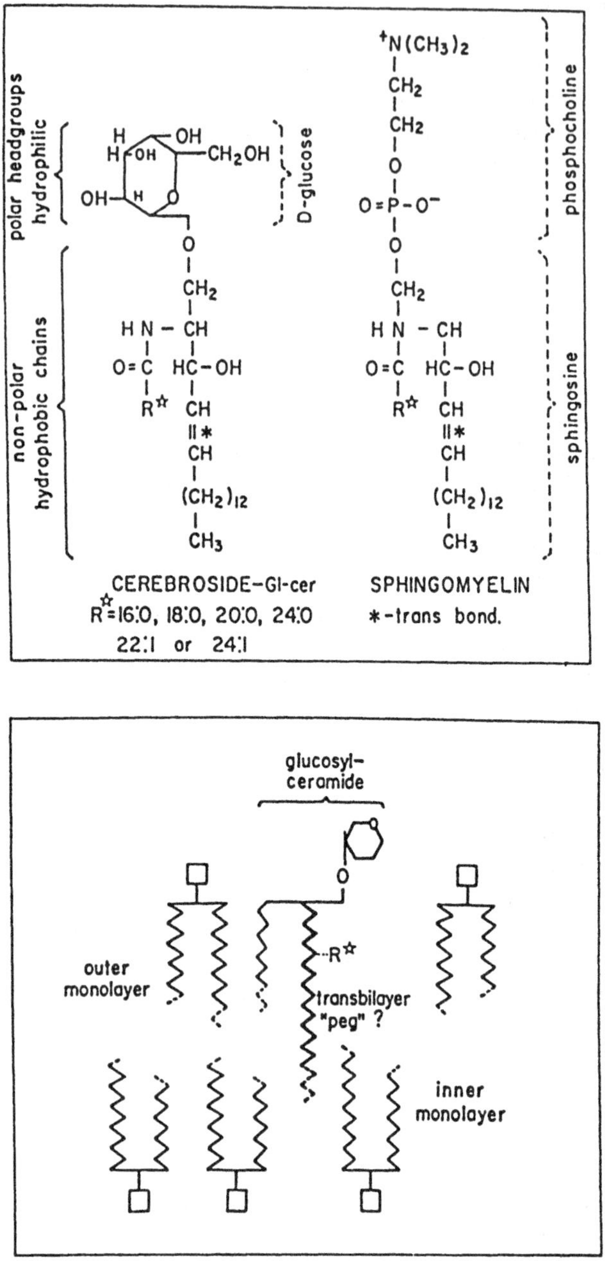

Fig. 3.8. Membrane Sphingolipids. D-glucose cerebroside is the second most common lipid category in the plasmalemma and tonoplast. *Above*: Like phospholipids, sphingolipids have a polar head and two non-polar tails one of which (R) is long chained fatty acid. The second polar tail is the 18C alcohol *sphingosine* or *dihydrosphingosine*. Part of this alcohol is derived from palmitoyl (16:0) Coenzyme A, while C atoms 1 and 2 of the sphingosine are derived from the β and α atoms, respectively, of the amino acid serine which forms the "hinge" of the molecule. Gangliosides, not depicted here, resemble cerebrosides but have multi-sugar headgroups some of which

metabolism of GlcCer in plant cells have involved several different approaches and are outlined below.

Compositional analyses have demonstrated that GlcCer from rye leaf plasma membrane contains primarily C_{22} and C_{24} saturated and mono-unsaturated hydroxy fatty acids. Hydroxysphingenine (*t*18:1) and isomers of sphingadienine (*d*18:2) are the predominant long chain bases. Analysis of intact underivatized GlcCer molecular species by HPLC indicated that the principal species in rye, *t*18:1, 24:1 GlcCer, comprises 40% wt of the total. Analyses of GlcCer from the leaves, roots and seeds of rye and other cereals as well as from dicotyledonous plants demonstrated tissue specific and genotypic differences with respect to acyl chain and long chain base composition. Typically, leaf and root GlcCer contain higher proportions of long chain (C_{22} and C_{24}) hydroxy fatty acids and trihydroxy long chain bases than do seed lipids. GlcCer from dicots have greater proportions of medium chain (C_{16} – C_{20}) hydroxy fatty acids.

Investigations of the phase properties of rye GlcCer have demonstrated that the lipid has a T_c (see pg. 121 for definition) of 56 °C and exhibits gel phase metastability. Individual GlcCer molecular species purified by HPLC also exhibit similar T_cs, but the metastable behavior differed depending on the degree of unsaturation of the constituent acyl chains.

3.5.2. *Sphingomyelins*

Sphingomyelins (Fig. 3.8) possess a phosphate upon which either *choline* or *ethanolamine* is found. These phosphocholine or phosphoethanolamine headgroups display a close resemblance to PC or PEA: sphingomyelins may thus also be classed under the phospholipid category. As in the case of PC and PEA the negative charge of the phosphate neck section is counterbalanced by the positive charge of the terminal alcohols – choline or ethanolamine – the final result being an *electroneutral* species. Sphingomyelins are present in most membranes of animal cells and especially in the myelin sheath of nerve cells. To date no reports of sphingomyelin occurrence in plants have been presented.

---

are conjugated to sialic acid. *Below*: Interdigitation of the very long chained fatty acids of plant sphingolipid cerebrosides (22:0, 22:1, 24:0 and 24:1) between the two individual monolayer leaflets lends stability to the bilayer and serves as a means of holding the two separate monolayers together.

3.5.3. *Gangliosides*

Like cerebrosides this sphingolipid has no phosphorus but does have a very large polar headgroup made up of several hexose sugar groups, some of which are conjugated to *sialic acid* which has a negative charge at pH 7.0. Six percent of the gray matter in the human brain is comprised of gangliosides which are believed to provide physiological receptor sites. As in the case of sphingomyelins there is no documentation of their presence in plant membranes.

While as seen above two of the three sphingolipid categories are apparently not present in plant membranes, or remain as yet undetected, the presence of relatively high levels of the first group – the cerebrosides – warrants a brief discussion of the biophysical effects in membranes of sphingolipids as a whole, and from which inference can be made to plant cerebrosides. This, therefore, is the topic of the next section.

3.5.4. *Biophysical effects of sphingolipids in membranes*

Sphingolipids as a group have a clearcut rigidifying effect on membranes and in this respect, after sterols, are considered a major agent contributing to membrane rigidification. This effect stems from the structural composition of the sphingolipid chains *both of which are saturated*, these being the long chained alcohol *sphingosine* or *dihydrosphingosine* (in which the double bond has been reduced) and the fatty acid chain which contains long chained species such as 16:0, 18:0, 22:0, 24:0 and some monounsaturated species – 22:1 and 24:1. Because of the specific nature and length of these chains the percentage of the non-polar hydrophilic section of sphingolipids is higher than that in phospholipids.

Another salient aspect of sphingolipid structure is that the fatty acyl tail is much longer than the sphingosine, and this staggered configuration may be instrumental in coupling bilayer leaflets, very much resembling a like mechanism in phospholipids where the *sn*-2 acyl chain is shorter than the *sn*-1 chain (see page 52). Fig. 3.8 (below) indicates how in a bilayer, the very long chained fatty acids may interdigitate between the two monolayer leaflets and so doing, add stability to the bilayer structure. Phase transition temperatures (T_c) of sphingolipids, depending on composition of the fatty acyl residue, are close to those of natural membranes. The amide bond (—NH—) is very prone to H^+ bonding and this may support proteins in the bilayer.

An interesting sphingolipid-associated trend has been observed in ageing and sensecence: in nature the sum of PC and sphingomyelin has been found to be constant, but with age relative proportions of each component change. In membranes of young tissues the main component of the left hand side of the equation

$$\text{PC} + \text{sphingomyelin} = \text{K}$$

is PC, while in ageing ones it is sphingomyelin, this going hand-in-hand – as reported in brain and nerve tissue – with increase of rigidity.

Accumulation of sphingolipids in heart tissue also decreases rate of "beating" of cardial cells, this being potentially reversed by PC. In keeping with this observation is the report, mentioned above, of Liljenberg's group (Norberg et al., 1991) that stress acclimation of oat plasma membranes *inter alia* is accompanied by a *decrease* in cerebroside and a parallel derigidification-associated decrease in sterol/free fatty acid ratio.

3.6. Chloroplast-specific glycerolipids

It has been found that chloroplast membranes, including those of the envelope, thylakoids and lamellae in the stroma, are very different from other cellular or organelle types. Considering the global coverage with vegetation and the latter's photosynthetic activity, Gounaris et al. (1986) have pointed out that the photosynthetic membrane probably is *the* most prevalent biological membrane just as the CO_2-fixing enzyme RuBP-carboxylase in C-3 plants is the most prevalent biological protein.

These plastid membranes are characterized by low content of phospholipids in general and especially by complete absence of PEA. These membranes are also typified by high contents of glycerolipids. It is customary to include membranes of proplastids, chromoplasts, leucoplasts, amyloplasts and etioplasts under the overall heading of "plastid-membranes" in addition to those of chloroplasts. Glycerolipids found in these are of two categories:

1. Galactolipids.
2. Sulpholipids.

3.6.a. *Galactolipids*

Fig. 3.9 shows the structures of the two major membrane galactolipids which according to Joyard and Douce (1987) comprise 2/3–3/4 of the

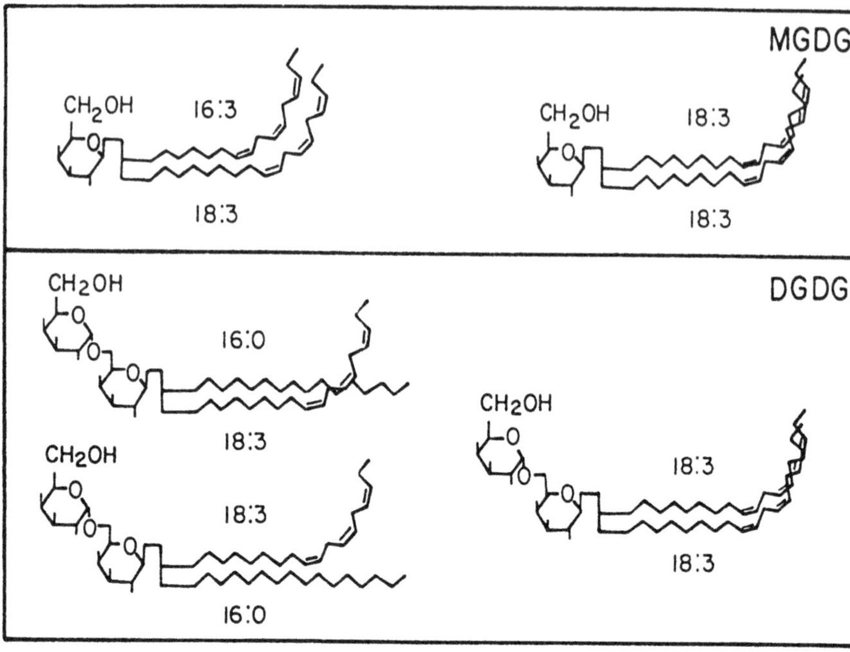

Fig. 3.9. Galactolipids of plant plastids and chloroplasts. In the headgroups of this category of membrane lipids the sugar D-galactose replaces the second alcohol which is found in phospholipids. The molecules drawn on the right represent the eukaryotic pathway of synthesis while those on the left, the prokaryotic pathway.

chloroplast envelope's lipid. It should be noted that due to the specific structure of the headgroups and in particular to the lack of a phosphate "neck" section, *galactolipids are electroneutral.* This property is of major importance in certain biophysical aspects which will be discussed in a later chapter. Moreover contrary to most other glycerolipids, when placed upon an aqueous solution MGDG does *not* assume a layered configuration but rather assembles into an *inverted cubic phase* which is believed to play a role in the curved sections of thylakoidal membranes.

Fig. 3.9 also indicates that if one galactose is present, the galactolipid is *monogalactosyldiacylglyceride* (MGDG) and if two galactose molecules are present the lipid formed is *digalactosyldiacylglyceride* (DGDG). In higher plant chloroplast membranes there is twice as much MGDG as DGDG while in algae the relative content of MGDG is even higher.

These two galactolipids also differ from the regular pattern of fatty acyl tailgroups as observed in phospholipids where as discussed earlier the *sn*-1 and -2 sites are respectively occupied by a long chained saturated and an unsaturated fatty acyl chain. Fig. 3.9 indicates that

Table 3.2. Pattern of conifer chloroplast membrane glycerolipid adaptation to frost hardening (according to Wolfenden and Wellburn, 1991)

Fatty acyl species	Frost hardened conifers	Others
18:2	Δ 5,9	Δ 6,9
18:3	Δ 5,9,12	Δ 6,9,12
18:4	Δ 5,9,12,15	Δ 6,9,12,15

here *both* chains may be unsaturated – containing 18:3 linolenic acid. Where a saturated residue occurs it is of interest to note that the fatty acyl species is *palmitic* acid (16:0) and not stearic acid (18:0) as often encountered in most other plant membranes.

With regard to the fatty acyl tail moieties of chloroplast membrane MGDG in coniferous tree needles, a unique situation exists. Work carried out in Wellburn's laboratory at the Institute of Environmental and Biological Sciences of the University of Lancaster, U.K., has shown that with increased adaptability to cold stress, the basic 18 C fragment undergoes progressive desaturation in manner generally unlike other 18 C species. This is manifest as diene, triene or tetraene (18:2, 18:3 and 18:4) occurrence respectively at the Δ5, 9; Δ5, 9, 12; or Δ5, 9, 12, 15 sites and *not* as ordinarily encountered at the Δ6, 9; Δ6, 9, 12 and Δ6, 9, 12, 15 sites. This is shown in Table 3.2.

The above research has shown that conifer chloroplast glycerolipid membranes of species able to withstand only a limited degree of freezing (ca −8°C) are less unsaturated than those able to withstand temperature of −18°C and below. It moreover has been shown that the industrial pollutant ozone, inhibits conifer needle adaptive response in this respect. These findings suggest a cold stress modulating mechanism achieved by maintaining membrane fluidity by two ways at least – (a) increase of unsaturation, (b) introduction of the initial C = C bond earlier into the fatty acyl chain – at C-5 instead of C-6. This could further increase degree of bending of the chain and hence also of membrane fluidity. These observations imply the appearance, under cold stress, of an as yet undetected C-5 desaturase enzyme.

Two different galactolipid metabolic pathways which may be listed in two separate groups function in plants:

1. *16:3 plants* – which follow a *prokaryotic* mode of biochemical synthesis in which galactolipids are synthesized in their entirety in plastids, and whose fatty acyl chains are 16C or 18C. These plants include clover, beans and the castor plant.
2. *18:3 plants* – which follow an *eukaryotic* mode of biochemical synthesis of galactolipids and in whose assembly both plastids and

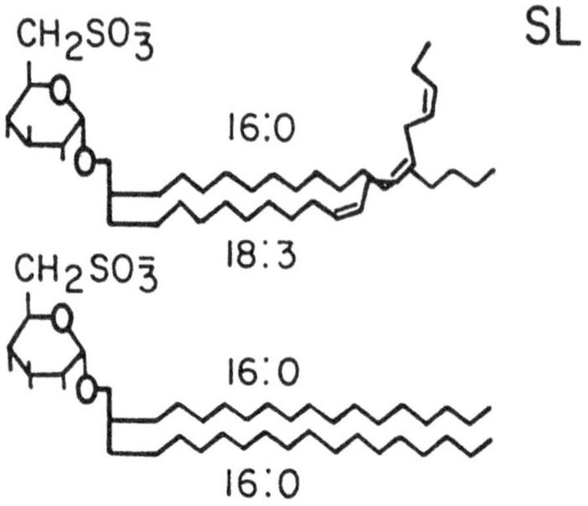

Fig. 3.10. The sulpholipid – sulphoquinovosyldiacylglycerol. This species is found in chloroplast membranes, and in animals in the central nervous system.

the endoplasmic reticulum participate. Such plants include spinach, tomatoes, *Hedera* (ivy) spp., irus and others.

The reader interested in detailed topographic distribution of fatty acyl residue in thylakoidal membranes of various plant species is referred to the systematic and well presented reviews of Unitt and Harwood (1982), Siegenthaler (1982) and Douce et al. (1987). These three reviews clearly indicate an asymmetrical distribution of MGDG and DGDG on the individual monolayer leaflets of the bilayer.

3.6.b. *Sulpholipids*

A particular type of glycerolipid designated *sulphoquinovosyldiacylglycerol* (SQDG) is found in the lamellae membranes of the chloroplast stroma. This glycerolipid somewhat resembles MGDG but differs in that the galactose headgroup has a sulpho-group attached to it (Fig. 3.10). As opposed to the electroneutral galactolipids, the addition of the sulphonic acid imposes a negative charge on SQDG thus making it markedly anionic and hydrophilic. It has been suggested that its presence in photosynthetic membranes affects availability of water for photosynthesis.

Concerning fatty acyl tailgroups, SQDG usually has palmitic acid (16:0) at the *sn*-1 site and linolenic acid (18:3) at the *sn*-2 site; however species of SQDG exist where both *sn*-1 and -2 have palmitic acid (cf. Fig. 3.10). Besides occurrence in chloroplast stroma, this sulpholipid is found in the central nervous system of some animals and in ants.

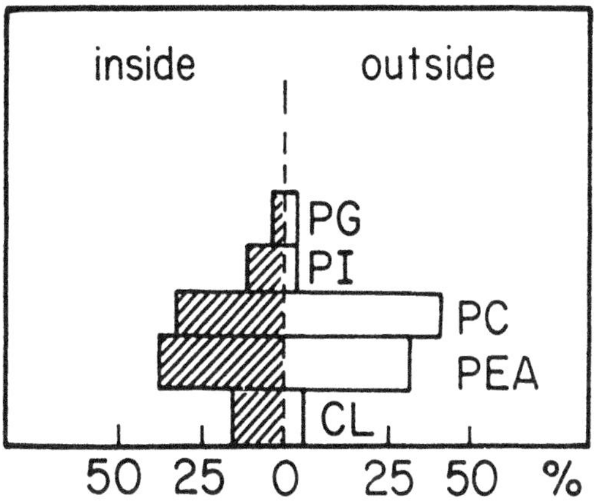

Fig. 3.11. The unequal distribution of phospholipids in the two bilayers comprising the "double double" – bilayered mitochondrial membrane (After Daum, 1985).

3.7. Mitochondrial phospholipids

The mitochondrial membrane resembles that of the chloroplast envelope in being a *double* double layer i.e. the mitochondrion is enclosed within *two* bilayer membranes, however this is as far as the resemblance goes. In terms of ontogenetic development, membrane phospholipid constituency in plant mitochondria is identical to that in human cardial mitochondria and, as in the plasmalemma, PC and PE are the major species. This resemblance, barring the high relative content of PC, also extends to mitochondrial membranes of procaryotic organisms such as *Cyanobacteria* and *Prochloron* which according to Douce et al. (1987) is the primordial ancestor of the chloroplast.

Fig. 3.11 indicates a typical analysis comparing phospholipid content of the inner to that of the outer bilayer of the plant mitochondrion. It can be noted that while all species found on the outer bilayer also occur in the inner one most of the phosphatidylinositol and cardiolipin is present in the inner bilayer.

As regards the glycerolipids, the mitochondrial membrane system is typified by a complete lack of galactolipids and sulpholipids.

3.7.1. *Cardiolipin (CL)*

This unique four-acyl-chained phospholipid is solely confined to mitochondria – especially to the inner membrane bilayer – where it is implicated in several physiological activities possibly associated with its exceptionally tight protein association. The existence of C18-C18

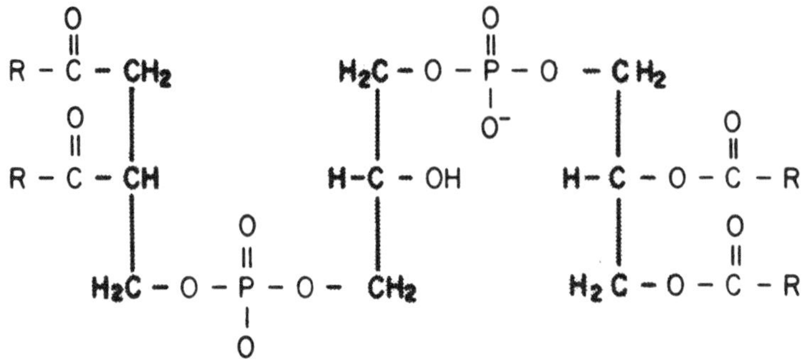

Fig. 3.12. Cardiolipin or diaphosphatidylglycerol DPD. Cardiolipin is mainly present in the mitochondrial inner membrane. Note that unlike the regular phospholipids this species has *four* fatty acyl tails, three glycerol residues and two phosphate "neck" sections.

double unsaturated moieties (18:2 linoleic acid) seems to be common in CLs from mammals, yeast and higher plants (Schlame and Beyer, 1991). Since CL possesses two sets of phosphate groups it is synonymously designated *diphosphatidylglycerol* (DPD). See Fig. 3.12. A similar phospholipid possessing only one phosphate group – *monophosphatidylclycerol* (MDP) is also present in plant mitochondrial membranes – apparently in higher amounts than DPD.

3.8. Platelet activating factor (PAF)

PAF is an unusual phospholipid with an ether linkage to a long chain alcohol at the *sn*-1 position, and an ester bond to acetic acid at the *sn*-2 position of the glycerol backbone while the polar *sn*-3 headgroup, like PC, is phosphocholine. Until recently this phospholipid was not detected in plants. In mammals it has long been known that PAF participates in blood clotting, and that it also contributes to the asthma syndrome by virtue of its marked ability to induce constriction of respiratory vessels (Barnes et al., 1988). Scherer et al. (1988) first isolated PAF in plants where, like in animalian systems, it appears to activate protein kinases or ATPase-driven proton pumps in the membrane. It thus transpires that PAF, which in reality is a PC species, while not being an integral membrane phospholipid component like regular PC, nevertheless is a regulatory agent acting on membrane-associated components. The putative role of PAF in signal transduction is discussed in Chapter 12.

Table 3.3 presents all the above-discussed membrane glycero- and sulpholipids in order of negative charges.

Table 3.3. Arrangement of plant membrane lipids with reference to their potential electrical charges. Full realization of anionic charge is pH dependent, high pH promoting full expression. (One negative charge in the phospholipids is from the phosphorous neck section, the others from headgroup moieties).

Lipid species	Net charge	Remarks
a. Electroneutral		
GlCer	0	completely electroneutral since headgroups lack electric charge
MGDG	0	
DGDG	0	
PC	0	Zwitterionic in nature, having one+, and one −, which cancel out.
PE	0	
PEA	0	also a Zwitterion, but + charge is on a primary NH_3, therefore tends to be more +
b. Anionic (negative)		
PG	−1	
PA	−1	
PS	−1	have two −, and one +
PAF	−1	
SQDG	−1	− charge from the SO_3^- in headgroup
PI	−1	
CL	−2	has 2 phosphate neck sections
PIP	−3	
PIP_2	−5	

3.9. The tertiary structure of membrane glycerolipids

The customary graphic representation of glycerolipid structure (employed in the present text as well) as seen for example in Fig. 3.1, conceals certain information since the impression is one of a flat molecule or of a molecule in which all parts are aligned along a single plane. However in reality it now has become apparent that they are *biplanar* – somewhat resembling the palm of a hand waving goodbye. As seen in Fig. 3.13, the polar hydrophilic headgroup and the phosphate neckgroup are aligned along one plane, while the non-polar fatty acid chains are aligned along a second plane which is at a right angle to the first one (Storch and Kleinfeld, 1985). Moreover carbon atoms #1 and 2 of the fatty acyl tail located at the *sn*-2 position have been found to be on the same plane as the headgroup and *not* in the plane of the major section of the fatty acyl tailgroups. In a three dimensional configuration, this particular location of the two initial C atoms of the *sn*-2 tail together with the bend induced by the cis

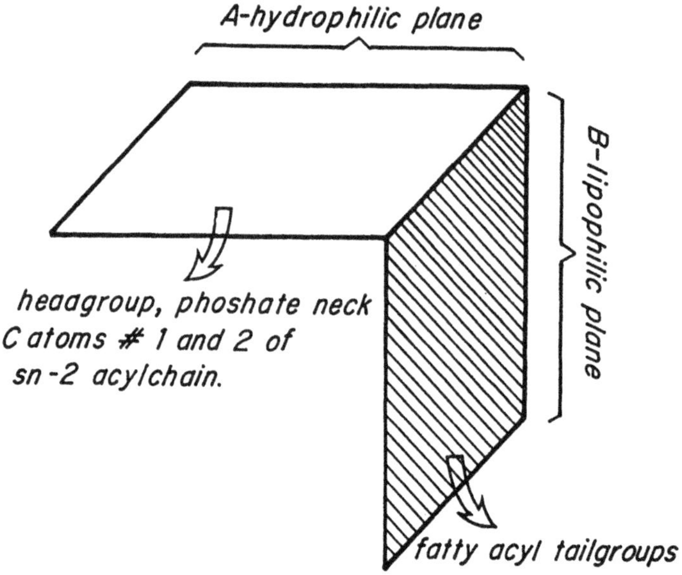

Fig. 3.13. The biplanar structure of glycerolipids. This diagram indicates that the headgroup is at right angles to the non-polar fatty acyl chains.

C=C bonds in this tail results in a situation in which even for the same number of C atoms, the *sn*-2 chain is considerably shorter than the *sn*-1 chain. Concomitantly, when located in bilayers, the fatty acyl chains which point inwards, with varying and unequal lengths of tailgroups, are believed (as in the similar case of the sphingolipid-cerebroside pointed out earlier in this chapter) to stabilize leaflet coupling of the bilayer.

In view of the above, in pictorial representation each monolayer leaflet may resemble a fleet of sailing boats at sea whose bodies representing the hydrophilic polar groups, are on the plane of the

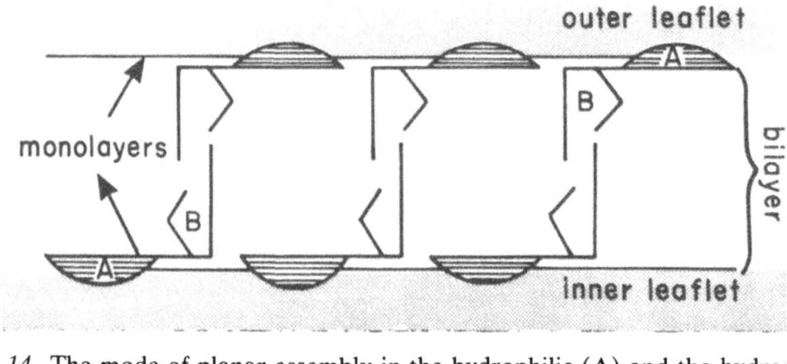

Fig. 3.14. The mode of planar assembly in the hydrophilic (A) and the hydrophobic (B) glycerolipid subunits in the bilayer.

aqueous medium (Fig. 3.14). Each boat may be envisaged to have two masts – one erect and the other bent. In a bilayer assembly the lipophilic mast sections may overlap since their heights are not equal (see Table 2.1 for actual dimensions).

3.10. References and recommended reading

Arondel, V., Tchang, F., Vergnolle, C., Grosbois, M., Guerbete, F., Jolliot, A., Morch, M., Pernollet, J., Delsemy, M., Puigodomenech, P., and Kader, J.C. 1989. Plant lipid transfer proteins. pp. 341–50. In: Eds. P. Biacs, K. Gruiz and T. Kremmer. *The Biological Role of Plant Lipids*. Adakemiai Kiado, Budapest, and Plenum Press, New York.

Barnes, P.J., Fan Chung, K. and Page, C.P. 1988. Inflammatory mediators and asthma. *Pharmacol. Rev*. 40:49–84.

Batty, R.S. and Slinkard, A.E. 1989. Relationship between phytic acid and cooking quality in lentils. *Can. Inst. Food Sci. Tech. Jour*. 22:137–42.

Beatrice, M.C., Palmer, J.W. and Pfeiffer, D.R. 1980. The relationship between mitochondrial membrane permeability, membrane potential and the retention of Ca^{2+} by mitochondria. *J. Biol. Chem*. 225:8863–71.

Bishop, W.R. and Bell R.M. 1985. Assembly of the ER phospholipid bilayer: the PC transporter. *Cell* 42:51–60.

Clark, K.A. and Goldsmith, M.H.M. 1987. Effect of surface and membrane potentials on IAA uptake and binding by Zucchini membrane vesicles. pp. 99–112. In: Ed. D. Klämbt. *Plant Hormone Receptors*. Nato ASI series Vol. H10. Springer Verlag, Berlin, Heidelberg.

Cooke, D.T. and Burden, R.S. 1990. Lipid modulation of plasma membrane bound ATPase. *Phys. Plant*. 78:153–9.

Daum, G. 1985. Lipids of mitochondria. *Biochem. Biophys. Acta* 882:1–42.

Davies, E. 1987. Action potentials as mulifunctional signals in plants: a unifying hypothesis to explain apparently disparate wound responses. *Plant, Cell, Environ*. 10:623–31.

Davies, E. 1990. Wound signals and translation. pp. 518–30. In: Eds. R. Pharis and S. Rood. *Plant Growth Substances, 1988*. Springer Verlag, Berlin.

de Gier, J., van Echteld, C.J.A., van der Steen, A.T.M., Noordam, P.C., Verkleij, A.J. and de Kruijff, B. 1982. Lipid organization and barrier functions of membranes. pp. 315-25. In: Eds. J.F.G.N. Wintermans and I.J.C. Kuiper. *Biochemistry and Metabolism of Plant Membranes*. Elsevier, Amsterdam.

de Kroon, A.I.P.M., Timmermans, J.W., Killian, J.A. and de Kruijff, B. 1990. The pH dependence of headgroup and acyl chain structure and dynamics of PS studied by H-NMR. *Chem. Physics Lip*. 54:33–42.

Douce, R., Alban, C., Bligny, R., Bloch, M.A., Covès, J., Dorne, A.J., Journet, E.P., Joyard, F., Neuberger, M. and Rebeille, F. 1987. Lipid distribution and synthesis within the plant cell. pp. 255–63. In: Eds. P.K. Stumpf, J.B. Mudd and W.D. Ness. *The Metabolism, Structure and Function of Plant Lipids*. Plenum Press, New York.

Douce, R. and Joyard, J. 1980. Plant galactolipids. pp. 321-62. In: Eds. P.K. Stumpf and E.E. Conn. *The Biochemistry of Plants*. Academic Press, New York.

Erdélyl, A.P. 1989. Enzymic modification of sunflower lecithin. pp. 455–8. In: Eds. K. Gruiz and T. Kremmer. *Biological Role of Plant Lipids*. Plenum Press, New York.

Felle, H. 1982. Effects of fusicoccin upon membrane potential, resistance and current-voltage characteristics in root hairs of *Sinapis alba*. *Plant Sci. Lett.* 25:219-25.

Geiger, B. 1985. Microfilament-membrane interaction. *Trends. Biochem. Sci.* 10:456–61.

Gounaris, K., Barber, J. and Harwood, J.L. 1986. The thylakoid membrane of higher plant chloroplasts. *Biochem. Jour.* 237:313–26.

Honig, B.H. 1986. Electrostatic interactions in membranes and proteins. *Ann. Rev. Biophys. and Biophys. Chem.* 15:163–93.

Joyard, J. and Douce, R. 1987. Galactolipid synthesis. pp. 215–73. In: Eds. P.R. Stumpf and E.E. Conn. *The Biochemistry of Plants. Vol. 9. Lipids: Structure and Function*. Academic Press, New York, London.

Kader, J.C. 1985. Lipid binding proteins in plants. *Chem. Physics Lipids*, 38: 51–62.

Kader, J.C., Douady, D. and Mazliak, P. 1982. Phospholipid transfer proteins. pp. 276–311. In: Eds. J.N. Hawthome and G.B. Ansell. *Phospholipids*. Elsevier, Amsterdam.

Kömer, L.E., Kjellbom, P., Larsson, C., and Møller, J.M. 1985. Surface properties of right side out plasma membrane vesicles isolated from barley roots and leaves. *Plant Physiol.* 79:72–9.

Landau, E.M. and Leshem, Y.Y. 1988. Biophysical interactions of membrane anionic phospholipids with sphingolipid, calcium and auxins. *J. Exp. Bot.* 39:551–9.

Leshem, Y.Y. 1987. Membrane phospholipid catabolism and Ca^{2+} activity in control of senescence. *Physiol. Plant.* 69:551–59.

Leshem, Y.Y. 1991. Evidence of the presence and mode of action of a membrane-associated plant phospholipase A_2. pp. 53–5. In: Eds. P. Quinn and J.L. Harwood. *Plant Lipid Biochemistry. Structure and Utilization*. Portland Press, London and Colchester.

Leshem, Y.Y., Halevy, A.H. and Frenkel, C. 1986. *Processes and Control of Plant Senescence*. pp. 71–9. Elsevier Press, Amsterdam.

Ling, G.N. 1969. A new model for the living cell. A summary of the theory and recent experimental evidence in its support. *Int. Rev. Cytol.* 26:1–61.

Low, M.G. 1987. Biochemistry of the glycosyl-phosphatidyl-inositol membrane protein anchors. *Biochem. J.* 244:1–13.

Lucy, J.A. 1970. The fusion of biological membranes. *Nature* 227:815–7.

Lynch, D.V. 1991. Glycosphingolipids of plant membranes. pp. 47–52. In: Eds. P. Quinn and J. Harwood. *Plant Lipid Biochemistry. Structure and Utilization*. Portland Press, London and Colchester.

Mansfield, T.A. and Atkinson, C.J. 1989. Calcium and abscisic acid. How do they control stomata in the intact plant. p. 50. In: Eds. H. Göring and P. Koffman. *Stomata '89. Colloquia Pflanzenphysiol.* Humbolt University, Berlin.

McAinsh, M.R., Brownlee, C. and Hetherington, A.M. 1990. Abscisic acid induced elevation of guard cell cytosolic Ca^{2+} precedes stomatal closure. *Nature* 343:186–8.

Møller, I.M. and Crane, F.L. 1990. Redox processes in the plasma membrane. pp. 93–126. In: Eds. C. Larsson and I.M. Møller. *The Plant Plasma Membrane*. Springer Verlag, Berlin-Heidelberg.

Møller, I.M. and Lin, W. 1986. Membrane bound NAD(P)H dehydrogenases in higher plant cells. *Ann. Rev. Plant. Physiol.* 37:309–34.

Morré, D.J., Brightman, A., Wang, J., Barr, R. and Crane, F.L. 1988. Roles for plasma membrane redox systems in cell growth. pp. 45–55. In: Eds. F.L. Crane, D.J. Morré and H. Low. *Plasma Membrane Oxidoreductases in Control of Animal and Plant Growth.* Plenum Publishers, New York.

Mudd, J.B. and Kleppinger, F.K. 1987. Sulpholipids. pp. 275–90. In: Eds. P.K. Stumpf and E.E. Conn. *The Biochemistry of Plants. Vol. 9. Lipids. Structure and Function.* Academic Press, New York.

Norberg, P., Nilsson, R. and Liljenberg, C. 1991. Alterations in lipid composition and phase behaviour of oat root plasma membranes after induction of dehydration tolerance. pp. 65–68. In: Eds. P. Quinn and J.L. Harwood. *Plant Lipid Biochemistry. Structure and Utilization.* Portland Press, London and Colchester.

Olson, E.N. 1988. Modification of proteins with covalent lipids. *Prog. Lip. Res.* 27:177–98.

Palmgren, M.J., Sommarin, M. Ulskov, P. and Jorgensen, P.J. 1988. Modulation of plasma membrane H^+ATPase from oat roots by lysophosphatidyl-choline free fatty acids and phospholiphase A_2. *Physiol. Plant* 74:11–19.

Rawyler, A., Meylan, M. and Siegenthaler, P.A. 1990. MGDG synthesis in intact chloroplasts and its relations with lipid asymmetry in thylakoid membranes. In: Eds. P. Quinn and J.L. Harwood. *Plant Lipid Biochemistry. Structure and Utilization.* Portland Press, London and Colchester (In press).

Reinhold, L., Kaplan, A. 1984. Membrane transport of sugars and amino acids. *An. Rev. Plant Phys.* 35:48–83.

Ricker, J., Spener, F. and Kader, J.C. 1985. A phospholipid transfer protein that binds long-chained fatty acids. *FEBS Lett.* 180:29–32.

Scherer, G.F.E., Martiny-Baron, G. and Stoffel, B. 1988. A new set of regulatory molecules in plants. A plant phospholipid similar to platelet-activating factor stimulates protein kinase and proton translocating ATPase in membrane vesicles. *Planta* 175:241–53.

Schlame, M. and Beyer, K. 1991. The molecular species of mitochondrial cardiolipin. pp. 17–19. In: Eds. P. Quinn and J.L. Harwood. *Plant Lipid Biochemistry. Structure and Utilization.* Portland Press, London and Colchester.

Schroeder, F. 1984. Role of membrane asymmetry in aging. *Neurobiol. Aging* 15:323–33.

Seignuret, M., Zachowski, A., Hermann, A. and Devaux, P. 1984. Asymmetric lipid fluidity in human erythrocyte membrane: new spin-label evidence. *Biochem.* 23:4271–5.

Siegenthaler, P.A. 1982. Transmembrane distribution and function of lipids in spinach thylakoid membranes: rationale of the enzymic modification method. pp. 351–8. In: Eds. J. Wintermans and P. Kuiper. *Biochemistry and Metabolism of Plant Lipids.* Elsevier, Amsterdam.

Stafford, R.E. and Dennis, E.A. 1988. Lysophospholipids as surfactants. *Colloids and Surfaces* 30:47–64.

Storch, J. and Kleinfeld, A.M. 1985. The lipid structure of biological membranes. *TIBS* 10:418–21.

Takashima, K., Watanabe, S., Yamada, M. and Mayima, G. 1987. The amino-acid sequence of the non-specific lipid transfer protein from germinated castor bean embryos. *Biochim. Biophys. Acta* 870:248–55.

Tasaka, Y., Nishida, J., Higashi, S., Beppu, T. and Murata, N. 1990. Fatty acid composition of phosphatidylglycerols in relation to chilling sensitivity of woody plants. *Plant Cell. Physiol.* 31:545–50.

Unitt, M.D. and Harwood, J.L. 1982. Lipid topography of thylakoid membranes. pp. 359–62. In: Eds. J. Wintermans and P. Kuiper. *Biochemistry and Metabolism of Plant Lipids*. Elsevier, Amsterdam.

van Breeman, R.B., Wheeler, J.J. and Boss, W.I. 1990. Identification of carrot inositol phospholipids by fast atom bombardment IUS. *Lipids* 25:328–34.

van Paridon, P.A., de Kruijff, B., Ouwerkerk, R. and Wirtz, E.W.A. 1986. Polyphosphoinositides undergo charge neutralization in the physiological pH range; a ^{31}P NMR study. *Biochim. Biophys. Acta* 877:216–9.

Weltzien, H.V. 1979. Cytolytic and membrane perturbing properties of lysophosphatidylcholine. *Biochim. Biophys. Acta* 559:259–87.

Wolfenden, J. and Wellburn, A.R. 1991. Effects of summer ozone on membrane lipid composition in Norway spruce (*Picea-Abies*). *New Phytol.* 118:323–30.

4. Sterols

Sterols (see Fig. 4.1), an important component both in plant and animal membranes, have an interesting biophysical effect on membrane properties. Like the glycerolipids, sterols are amphiphilic and under low molar concentrations may form micelles. However in contrast to glycerolipids the hydrophilicity of the hydrophilic section in the sterols (the single -OH group) is less pronounced. Thus even at higher molar concentrations they do not form bilayers when overlaid on an aqueous surface. One of the marked effects of sterols in biological membranes is the *condensing effect*: as found by Demel and de Kruijff (1976) in the Netherlands the surface area occupied by an artificial membrane composed of phospholipid and cholesterol is *less* than the sum of areas of each component. The concept of a condensing action of sterols in membranes is widely held, but evidence has recently been presented that this may not necessarily be so since methods utilized to establish this effect purportedly are not applicable to biological membranes (Matubaya et al., 1989).

4.1. Structure and function

The molecular alignment of a sterol in a biological membrane is such that its polar group – the hydroxyl (—OH) section – is associated with the polar phospholipid headgroups. This —OH group appears to be essential for the inclusion of the sterol molecule as a whole into the membrane since if for example it is substituted by a Cl_3 group, sterol association with the membrane is obstructed. The sterol's cyclopentanoperhydrophenantrene skeleton comprised of the A, B, C and D rings and the aliphatic tail are arrayed parallel to the fatty acyl tailgroups of the membrane phospholipid. It is presently not quite clear how this non-polar hydrophobic section of the sterol molecule associ-

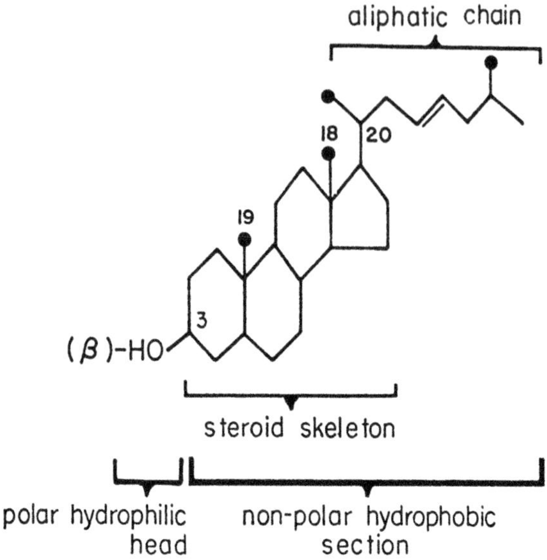

Fig. 4.1. The structure of stigmasterol – a typical plant sterol. The mammalian cholesterol differs from stigmasterol only by the lack of an unsaturated double bond (—C═C—) in the aliphatic sidechain. As here depicted the plane of the —OH group faces upwards, at right angles to the plane of the steroid skeleton, this configuration being termed β-OH, and being the normal biological form. If the —OH group faces downwards the resulting configuration termed α-OH is biologically inactive. ⫯ indicates methyl groups facing the same way as the —OH.

ates with these phospholipid fatty acyl tails. It has been surmised that the methyl group at positions 18 and 19 associate with, or enter stereochemical "clefts" formed by the *cis* C═C bonds present in the polyunsaturated fatty acyl chains (18:2, 18:3, etc.) which are usually present between carbons 9 and 10, 11 and 12, etc. in membrane fatty acyl species. This assumed interaction is schematically depicted as follows:

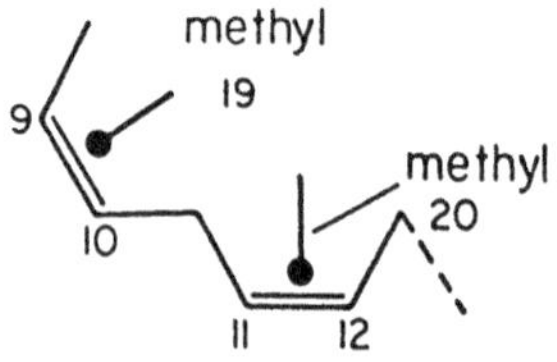

Since the steroid skeleton is rigidly planar, sterols on the one hand cause membranes to be more rigid or, in other words, they increase microviscosity. On the other hand, the hydrophobic and aliphatic sterol tail which inherently is mobile introduces a certain degree of disorder in the microdomain of the membrane with which it is associated – it thus has an opposite and fluidizing effect.

Taken together the above findings indicate that membrane sterols have a homeostatic and stabilizing effect which enables cellular function over a wide temperature range including diurnal ambient temperatures which may fluctuate to the extent of 30°–40°C. Were sterol not present in membranes, microviscosity changes would be biologically lethal in terms of trans-membrane metabolite exchange, loss of fluidity, "leak" formation etc. The effect of sterols on membrane phase change is dealt with in Chapter 7, and the important aspect of membrane sterol contribution to the ability of plants to withstand chilling and cold stress is discussed in detail in Chapter 10.

4.2. Occurrence and distribution

As pointed out by John Goad of the University of Liverpool, the major sterols found in higher plants, algae, fungi and many protozoa are typically substituted at C-24 (see Fig. 4.1) in the side chain by a methyl or an ethyl group (Goad et al., 1988). Depending on the organism, the C-24 alkyl sterol can have either the 24-R or 24-S configuration and the extra carbons are derived from transmethylation reactions. Why plants, fungi and protozoa should expend considerable energy to synthesize 24-alkyl sterols is an intriguing question and despite much speculation there is as yet no satisfactory answer.

Goad and workers have established that sterols fulfill two functions in plant cells and may be classified as follows:

(a) *Bulk sterols*, associated with membrane architecture – possibly acting as outlined in the preceding section.

(b) "*Trigger*" *sterols* – required in trace amounts but nevertheless being an obligatory requirement for cell proliferation.

According to these workers the "bulk" role is best fulfilled by 3-β-hydroxyl-Δ^5 sterol, while the structure of the side chain is not crucial. By contrast it is apparent that sterols with a 24 *ethyl* group are more effective than 24 *methyl* sterols (Tal and Nes, 1987).

Cooper and Strauss (1985) have indicated that sterols are not present in all types of cellular membranes. It is customary to assign sterols a more prominent role in membrane physiology of eukaryotic nature and less so in organelles of prokaryotic origin. Following this line of reasoning, sterols are presumably more prevalent and play an important role in the plasma membrane and far less so in mitochondrial membranes. In the plant plasmalemma 5–10% of the total sterol is not sterol *per* se but rather *sterolglycoside*. In keeping with this, in chloroplast thylakoid membranes only small amounts of sterols (0.2–0.6%) have been detected; however even this small amount appears to be important for electron transfer since antiserum to sitosterol when applied to tobacco, spinach and *Delphinium* chloroplasts inhibited the transfer. In mammalian ER membranes there is very little sterol present, and what is present is found in both leaflets of the bilayer and as contrasted to glycerolipids, ER sterols flip-flop quite easily. A parallel may or may not exist in plant ER membranes: information on this aspect of plant ER sterols is still lacking.

4.3. Plant membrane sterols

The major plant membrane sterols are *sitosterol*, *campesterol* and *stigmasterol*. Unusual sterols are found in some species e.g. *spinasterol* in spinach and *cycloartenol* in oat plasma membranes. Despite the belief that this is not the case, *cholesterol* – the major human sterol – *does* occur in plants albeit in minute quantities. Garg et al. (1987) working on both monocots and dicots have indicated that at certain developmental stages meristem primordia may contain 15–45% cholesterol. The reader, who after having read the former sentence, now fears for his heart may rest assured since meristem primordia comprise only a very small percentage of total tissue from which they were excised, and cholesterol if at all present, comprises an even smaller concentration in edible tissues. Moreover what is to be feared is *cholesterolester* and not free cholesterol. The latter has also been found in minute quantities in date palm pollen and in cultured celery (*Apium graveolens*) cells.

The above sterols and some others in their natural form are the major membrane sterol categories, but in plant membranes approximately 5–10% may be conjugated to glucose to form *sterolglycosides* or as sometimes termed in full: *acylsterolglycosides* (= sterol + a fatty acyl chain + glucose).

In pathogenic fungal membranes including those of *Phytophtera* (potato blight) somewhat different sterol species such as *lanosterol* may appear. In view of potential agricultural application to prevent fungal infection (Parish et al., 1987), a promising line of research is the synthesis of specific inhibitors of lanosterol biosynthesis such as *imino-squaline*.

Growth inhibitors of the triazole group such as *paclobutrazol* which interferes with gibberellin biosynthesis and curbs vegetative growth

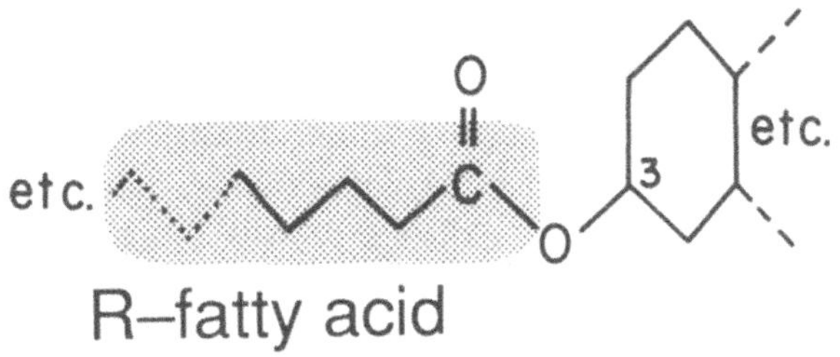

Fig. 4.2. Esterification of the polar OH group by a fatty acyl residue to form a sterol-ester. Cholesterol-ester is a blood vessel-blocking agent. R = a long-chained fatty acyl residue e.g. oleic acid.

have also been found to effect sterol biosynthesis. In celery and yeast cultures Haughan et al. (1989) have reported that paclobutrazol blocks the biosynthesis of sterols at the α-methylation stage of carbon atom #14 and that as a result there is an accumulation of shorter sterols possessing only 14 C atoms e.g. *obtusifoliol*. A further effect of paclobutrazol *is to cause a marked decrease of membrane phospholipid unsaturation* – levels of linoleic acid (18:2) decreasing, this being accompanied by an increase in levels of palmitic (16:0) and of oleic acid (18:1).

4.4. Sterol esters and heart disease

The material presented in the preceding sections of the present chapter apparently indicate that the presence of sterols (including cholesterol) in biological membranes is desirable in that they contribute to optimal cell function and possess a stabilizing effect on membrane microviscosity under vicissitudes of macro- and microenvironmental changes. If this were indeed so then a question may be posed as to extent of damage caused in *atherosclerosis* – a condition of arterial or venous blockage leading to cardial disorders or to stroke. Decisive research conducted by M. Shinitzky and co-workers (1984) at the Weizmann Institute in Rehovoth and by others has indicated that in this respect a clear distinction should be made between the so called "beneficial cholesterol" which is membrane-located and acts as outlined above, and between *cholesterol-esters* such as *cholesteryl-oleate* in which the polar —OH group at C atom #3 undergoes esterification with a long chained fatty acid residue (see Fig. 4.2). These sterol-esters *and not the cholesterol* precipitate on blood vessel wall lining and are the cause of the above mentioned maladies. Hence according to Shinitzky, blood testing for cholesterol should present both parameters – free cholesterol and cholesterol-ester.

According to Mead et al. (1986) human diet in developed countries averages the daily consumption of 200–300 mg plant sterols. These are only minimally ingested in the digestive tract, and what does penetrate is unesterified. When consumed in greater amounts, the above workers found that the plant sterols prevent the absorbtion of animal cholesterol in the digestive system since they compete with cholesterol for entry via the mucous membranes and also inhibit cholesterol esterifcation.

Sterol esters have been identified in plants but their function has not been defined. Their characterization has revealed a selectivity of esterification. Working on cell suspensions of celery (*Apium graveolens*) Dyas et al. (1991) have reported that whereas the free sterol mixture consists almost exclusively of campesterol, stigmasterol, sitosterol and isofucosterol, the steryl esters include a large proportion of esterified *precursor sterols*, most notably *cycloartenyl-linoleate*. This implies that the esterified fatty acid is exclusively linoleic acid (18:2). How well the celery pattern reflects the overall one in higher plants requires further research.

We mention that the *lipoproteins* present in mammalian blood serum – chylomicrons, very low density lipoprotein (VLDL), intermediary density lipoproteins (IDL) and high density lipoproteins (HDL) – which serve to transfer cholesterol to various parts of the body, have not been found in plants. There however is a certain degree of similarity to the "phospholipid transfer proteins" mentioned in Chapter 5, but the analogy should not be drawn too far since the animalian lipoproteins mentioned above are serum-located, while the plant phospholipid transfer proteins are intercellular.

4.5. Phospholipid/sterol ratio as a senescence index

Within a given botanical species and membrane category, in an intact and active cell possessing functional membranes, the phospholipid/sterol ratio is constant. With onset of senescence or upon exposure to stress, the component undergoing the most pronounced change or breakdown is the phospholipid whereas the sterol is more stable and far less prone to degradation. Thus in a senescencing membrane the phospholipid/sterol ratio decreases, and if this ratio in the intact membrane is determined, that of senescencing cells could be employed as an index reflecting degree of senescence or conversely – of cell integrity. An alternative means of expressing what essentially is the same parameter is the *free fatty acid/sterol* ratio and not the phospholipid/sterol one. The rationale underlying this approach is that phospholipid catabolism involves release of the fatty acyl tails – the free fatty acid thus obtained acting as a detergent which causes further disruption of the membrane. However in this case the ratio *increases* with senescence in contrast to the decrease of the former. Employing this parameter in either of its forms (phospholipid/sterol or phospholipid/free fatty acid ratios), changes with senescence of flower petals (Thompson et al., 1982; Borochov and Faiman-Weinberg, 1984) and of pea foliage (Sridhara and Leshem, 1986) as well as with exposure to low temperatures (Makhlouff et al., 1990) have been reported.

It is more than of passing interest to mention that when plants become acclimated to environmental stress such as cold, an *opposite* trend is noted i.e. more phospholipids relative to sterols are synthesized. Typical cases are those reported in the plasma membranes of living bark cells in the mulberry and of oat roots (Yoshida, 1984; Norberg et al., 1991).

4.6. References and recommended reading

Borochov, A. and Faiman-Weinberg, R. 1984. Biochemical and biophysical changes in plant protoplasmic membranes during senescence. *What's New in Plant Phys.* 15:1–4.

Cooper, R.A. and Strauss, J.F. 1985. Regulation of cell membrane cholesterol. pp. 74–98. In: Ed. M. Shinitzky. *Physiology of Membrane Fluidity. Vol. 1.* CRC Press, Boca Raton.

Demel, R.A. and de Kruijff. 1976. The function of sterols in membranes. *Biochim. Biophys. Acta* 457:109–52.

Dyas, L., Evershed, R.P., Prescott, M.C. and Goad, L.J. 1991. Steryl esters in a cell suspension culture of *Apium graveolens*. pp. 332–4. In: Eds. P. Quinn and J.L. Harwood. *Plant Lipid Biochemistry. Structure and Utilization.* Portland Press, London and Colchester.

Garg, V.K., Douglas, T.J. and Paleg, L.G. 1987. Presence of unusually high levels of cholesterol in the shoot apices of flowering plants. pp. 83–5. In: Eds. P.K. Strumpf, J.B. Mudd and W.D. Ness. *The Metabolism, Structure and Function of Plant Lipids.* Plenum Press, 724 pp.

Goad, L.J., Haughan, P.A. and Lenton, J.R. 1988. Regulation of sterol production and the effects on plant cell growth. *Brit. Pl. Gr. Reg. Group Monograph* 17:91–105.

Guye, M.G. 1987. Chilling and age related changes in the free sterol composition of *Phaseolus vulgaris* primary leaves. *Plant Sci.* **53**: 209–213.

Hartman, M.A. 1984. Sterol glucosylation in plasma membranes from maize coleoptiles, pp. 315–18. In: Eds. P.A. Siegenthaler and W. Eichenberger. *Structure, Function and Metabolism of Plant Lipids*. Elsevier, Amsterdam.

Haughan, P.A., Rolph, C.E., Lenton, J.R. and Goad, L.J. 1989. The effects of paclobutrazol on sterol and acyl lipid metabolism in celery and yeast. pp. 239–40. In: Eds. P. Biacs, K. Gruiz and T. Kremmer. *The Biological Role of Plant Lipids*. Akademiai Klado, Budapest, and Plenum Press, New York.

Makhlouff, J., Willemot, R., Couture, R., Ariel, J. and Castaigne, F. 1990. Effect of low temperature and controlled atmosphere storage on the membrane lipid composion of broccoli flower buds. *Scient. Hort.* 42:9–20.

Matubaya, N., Matsunaya, R. and Motomura, K. 1989. Interaction of cholesterol and octodecanol in a mixed adsorbed film at carbon tetrachloride/water interface: criticism about the condensing effect of cholesterol. *Langmuir* 5:1048–51.

Mead, J.F., Alfin-Slater, R.B., Howton, D.R. and Popjak, G. 1986. *Lipids: Chemistry, Biochemistry and Nutrition*. Plenum Press, New York.

Norberg, P., Nilsson, R. and Liljenberg, C. 1991. Alterations in lipid composition and phase behaviour of oat root plasma membranes after induction of dehydration tolerance. pp. 65–8. In: Eds. P. Quinn and J.L. Harwood. *Plant Lipid Biochemistry. Structure and Utilization*. Portland Press, London and Colchester.

Parish, E.J., Hanners, P.K. and Nes, W.D. 1987. *Synthesis and Biological Evaluation of Fungal Bioregulators of Sterol Biosynthesis*. pp. 103–5. *op. cit.* as Garg et al. above.

Shinitzky, M. 1984. The membrane fluidity and cellular functions. pp. 1–51. In: Ed. M. Shinitzky. *Physiology of Membrane Fluidity. Vol. 1*. CRC Press, Boca Raton.

Sridhara, S. and Leshem, Y. 1986. Phospholipid catabolism and senescence of pea foliage membranes. Parameters of Ca^{2+} calmodulin: phospholipase A_2 induced changes. *New Phytol.* 103:5–16.

Tal, E. and Nes, W.D. 1987. Regulation of sterol biosynthesis of sunflower suspension cultures. *Plant. Physiol. Suppl.* 83:161.

Thompson, J.E., Mayak, S., Shinitzky, M. and Halevy, A.H. 1982. Acceleration of membrane senescence in cut carnation flowers by treatment with ethylene. *Plant Physiol* 69:859–63.

Wang, S.Y., Sun, T., Whitaker, B.D. and Faust, M. 1988. Effect of paclobutrazol on membrane lipids in apple seedlings. *Physiol. Plant* 73:560–4.

Yoshida, S. 1984. Chemical and biophysical changes in the plasma membrane during cold acclimation of mulberry bark cells. *Plant Physiol.* 76:257–65.

5. Membrane proteins

The three major classes of membrane proteins are as follows:

a. *Peripheral or extrinsic proteins*. These are associated with only one leaflet – either the outer or the inner – of the bilayer.

b. *Integral or intrinsic proteins*. These are associated with both bilayer leaflets and may traverse the membrane more than once.

c. *Transport proteins*. These may coincide with either 'a' or 'b' above and also include proteins which while interacting with the membrane, are not anchored in it and may traffic in the cytosol from one membrane type to the other. Transport proteins can be grouped into the following three general classes:
   1. Pumps
   2. Carriers
   3. Channels

This classification is based on different forms of energy input: chemical bond energy (i.e. ATP) or light energy for pumps and the energy of electrochemical ion gradients for carriers (symport or antiport – more about which, later) and channels (uniport). Another basis for these categories is the speed of transport, i.e., the number of solute molecules transported per protein per second, with pumps relatively slow (less than 500 per second), carriers – intermediate (500 to 10,000 per second), and channels, the most rapid (10,000 to many million per second) (Sussman and Harper, 1989).

5.1. The peripheral proteins

These proteins are easily isolated by altering ionic strength or pH of the encasing medium. Since they are *not* associated with the internal hydrophobic bilayer reaction which plays an important role in anchoring integral proteins, peripheral protein:membrane association is mainly due to effects exerted by the *headgroups* of supporting glycerolipids and in particular of anionic species of the latter. Peripheral proteins include part of the membrane's enzymes, phytohormone receptors, coated pits, polyamines and sections of lectins. Divalent cation-mediated *electrostatic bonding*, outlined in detail in the next section, plays an important role in the association of peripheral proteins with membranes.

The aforesaid in no way excludes the importance of electrostatic crosslinking in anchoring of integral proteins in the bilayer: however in the latter protein category, it is believed that the major factor in this respect is the affinity of hydrophobic amino acids for fatty acyl sidechains in the membrane's interior as well as protein:protein interaction between integral and peripheral proteins.

5.2. The integral proteins

The proteins of this category may cross the membrane once or several times and are linked, either electrostatically or by means of biophysical lipophilicity, to the inner domains of the bilayer. Most integral proteins of membranes have a similar and unique structural motif: they are anchored in the membrane bilayer by one or more transmembrane stretches of 15–25 mostly hydrophobic amino acid residues. It is believed that each of these hydrophobic sequences spans the membrane as a single α-helix. This was first revealed by low-resolution structural analysis of bacteriorhodopsin, which showed seven transmembrane helical stretches per molecule, and was later confirmed by high resolution X-ray crystallographic analysis of the bacterial photoreaction center.

How do these hydrophobic sections become inserted into a membrane so that a particular integral protein molecule is correctly oriented in the membrane? It appears that such insertion is not spontaneous and does not occur directly into the lipid of the bilayer but requires energy and an organized system of proteins. Singer and Yaffe (1990) employing chloroplast, mitochondrial and ER membranes suggest that such integral protein insertion is aided by "*translocator proteins*". The translocator molecule is postulated to consist of an aggregate of *n* homologous transmembrane subunits forming a water-filled transmembrane channel down the central axis of the aggregate, similar to the structure of the acetylcholine receptor (cf. Fig. 5.9). A hydrophobic interface between two of the *n* subunits within the membrane serves as the passageway for the polypeptide to be translocated. It is

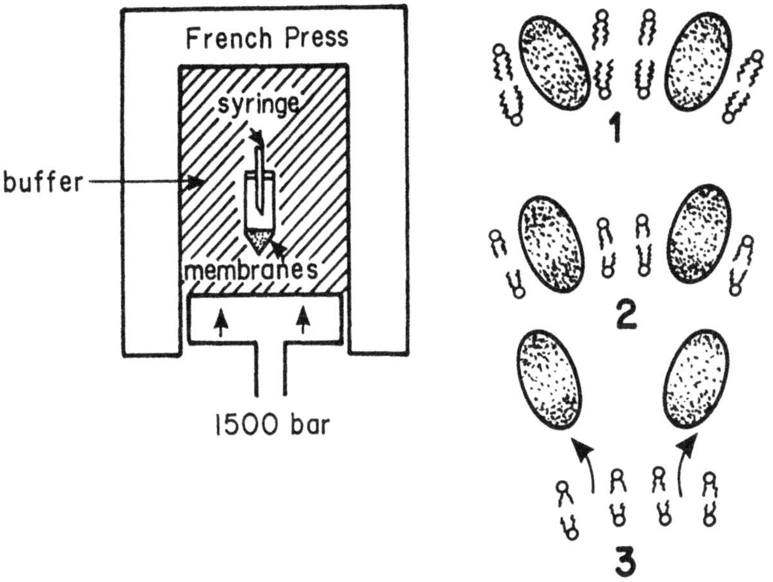

Fig. 5.1. The piezometric (pressure-associated) method for gentle extrusion of membrane proteins. This method is based on subjecting membranes to hydrostatic pressure (after Deckman et al., 1985).

surmised that the protein to be inserted possesses a specific N-terminal signal sequence (a "leader peptide") which initially associates with a *signal recognition* particle in the membranes.

By use of weak detergents such as Triton-100 which substitute for the fatty acyl domains of the bilayer it is possible to elute integral proteins without impairing their biological activity. Another method – developed for mammalian tissues – which is regarded as more "gentle" in terms of retaining protein integrity is by subjecting membranes to hydrostatic pressure in a French Press (Deckman et al., 1985). This method has been applied to plant membranes for the extrusion of membrane proteins associated with ethylene formation (Leshem, 1987). See also Fig. 5.1.

Integral proteins may be structurally grouped into two groups:

1. Simple integral proteins
2. Complex integral proteins

5.2.a. *Simple integral proteins*

These possess classic α-helical structure and traverse the membrane *only once*. It has been established that the helical sections which are located within the membrane have hydrophobic (or conversely –

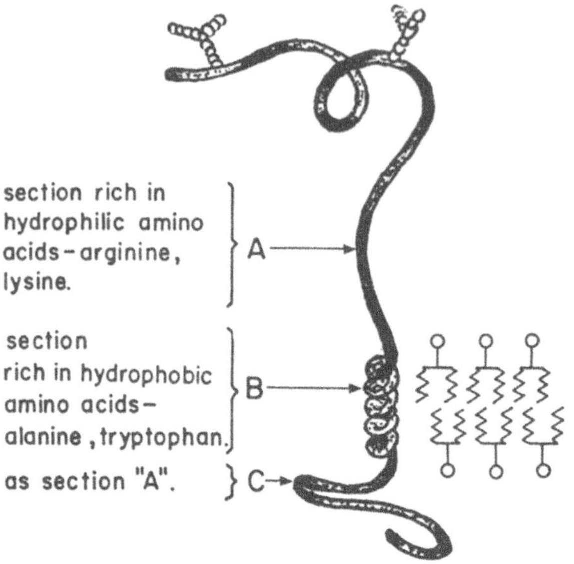

Fig. 5.2. Mode of association of a simple integral protein in a membrane.

lipophilic) amino acid residues in certain sections along the coil - these in turn having a non-covalent affinity for the fatty acyl sidechains of the membrane lipids. It moreover has been indicated that the length of this internal protein segment corresponds to the width of the bilayer. In contrast, the hydrophilic section of these simple proteins which faces outwards contains amino acid segments comprised primarily of hydrophilic residues e.g. arginine and lysine (Reithmeier, 1985). The above principles are illustrated in Fig. 5.2.

Schatz (1986) has summarized experimental evidence indicating that the following biophysical principles govern orientation of proteins in biological bilayers – these rules applying to both simple and complex proteins:

(a) A stable transmembrane anchor sequence contains an uninterrupted stretch of at least 16–17 uncharged amino acid residues. While this stretch may include hydrophilic residues (e.g. serine, threonine, cysteine, etc.), the stretch must have a high overall hydrophobicity.

(b) Artificially created transmembrane sequences may be as short as 11 residues or include a charged residue but are then less stable or may even be toxic to the cell, since the bilayer structure is perturbed. Therefore there is no sharp distinction between a transmembrane anchor sequence and a sequence that fails to anchor a protein in a membrane.

(c) Neither C-terminally-located basic residues, nor specific adjacent sequences are required for an anchoring sequence to function.
(d) No specific amino acid sequence motif is required, completely artificial sequences also function. However for ion channel specificity, sequencing apparently does play a role.
(e) Shifting the position of a transmembrane anchoring sequence in a protein alters the transmembrane disposition of the resulting protein in a predictable manner.

While the above-described mechanisms probably account for most simple integral membrane anchoring, Waneck et al. (1988) have demonstrated that some membrane protein may be covalently linked to the bilayer via phosphatidylinositol. Typical of such proteins is a short hydrophobic section within the membrane and a limited or even complete lack of protrusion into the cytosol or the membrane's exterior.

At this stage we stress a point, often overlooked by plant scientists, that most membrane proteins have an *oligosaccharide* tail on the externally facing side. The exact function of most of these tails in plant membrane physiology has not been elucidated.

A *proteoglycan*, which is a member of the large class of *arabinogalactan-proteins* has been isolated from the outer surface of plant plasma membranes. The assumed function of this protcoglycan is to provide cell surface attachment sites for cell wall matrix polysaccharides (Pennel et al., 1989); however the observation that such arabinogalacton proteins are also found albeit to a lesser extent on tonoplasts, indicates further and as yet unfathomed roles not connected with cell wall attachment.

These oligosaccharides together with other membrane carbohydrates such as glucose and galactose contained in galactolipids and sulpholipids and arabinogalactans may comprise approximately 20% of the plant plasma membrane as compared to the 8% carbohydrate of the human erythrocyte membrane (Larsson et al., 1990).

5.2.b. *Complex integral proteins*

In contrast to the α-helical simple integral proteins, the complex ones are globular – being comprised of several α-helical loops which may span the membrane several times in what is termed a *multiple hairpin bending* fashion. Membrane proteins belonging to this group include enzymes, ion channels, hormone receptors and possibly other proteins as well. An enzyme possessing one such loop is *cytochrome* b_5, while

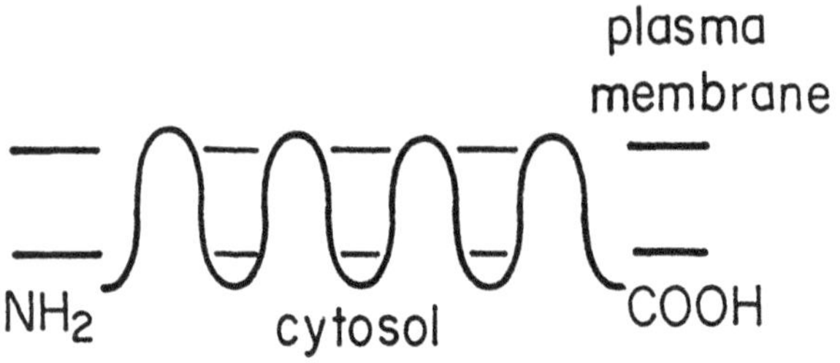

Fig. 5.3. Multiple looping of a complex integral protein in a membrane. Several of the ATPase family of enzymes adopt this configuration.

Ca^{2+}ATPase and Na/K ATPase spans the membrane several times. In *Arabidopsis* plasma membrane the H^{+}ATPase, similar to the same enzyme in the lower organism *Neurospora*, has four loops (as seen in Fig. 5.3): 20% of its protein is embedded in the membrane and is located in the eight transmembrane segments, 70% is hydrophobic and exposed on the cytosolic interior surface, while only 5% is exposed to the extra-cellular surface (Harper et al., 1989).

5.3. Ca^{2+} mediated electrostatic bridging of complex integral proteins

In addition to the complex of factors which serve to hold integral globular proteins in the bilayer, those which serve to anchor the thread-like simple proteins apparently differ somewhat. Their intra-membranal protein segments are less hydrophobic than corresponding segments of simple integral proteins. Hence the greater relative importance of electrostatic bridging mediated by divalent cations such as Ca^{2+} and Mg^{2+} which contribute to a greater extent to their stationing within the bilayer.

Membranal stability, microviscosity (η) and bulk lipid phase state may be controlled by Ca^{2+}. All negatively charged phospholipids such as phosphatidylinositides or phosphatidylserine, by virtue of electrostatic crosslinking with the $^{-}$O—P—OH section(s) of the headgroup induce physiologically meaningful structural rearrangements in membranal architecture. Leshem (1987) has shown that Ca^{2+}-mediated electrostatic crosslinking of proteins to membrane glycerolipid moieties includes the following possibilities (See Fig. 5.4):

(1) A multiple phospholipid-phospholipid bonding between phospholipids themselves inducing partial separation of charged phospho-

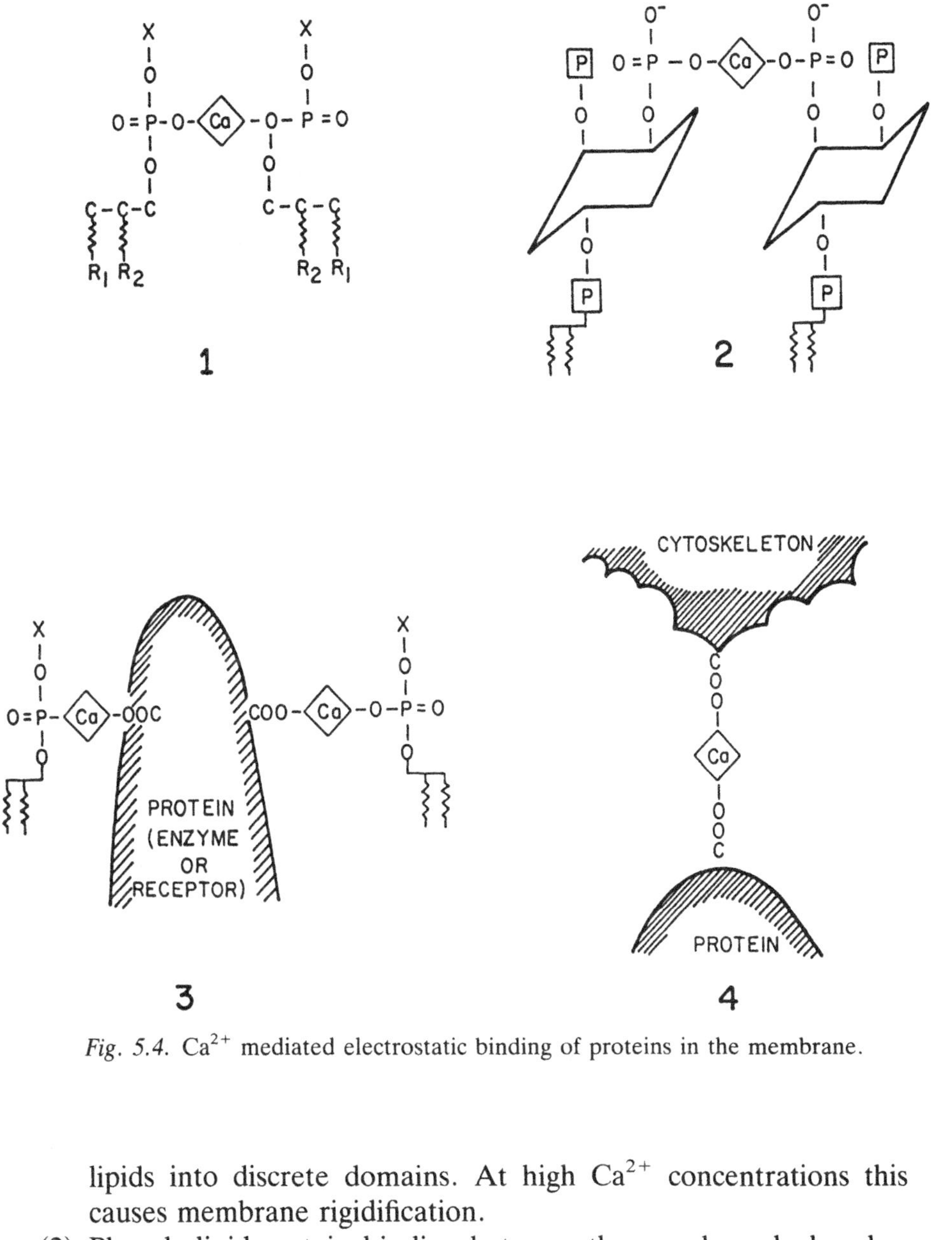

Fig. 5.4. Ca^{2+} mediated electrostatic binding of proteins in the membrane.

lipids into discrete domains. At high Ca^{2+} concentrations this causes membrane rigidification.

(2) Phospholipid-protein binding between the membranal phospholipids and the carboxyl tails of membrane proteins which restricts protein motility.

(3) A protein-cytoskeleton bridging of protein chemical nature as in '2', or an indirect effect promoting protein-cytoskeleton binding via the γ-glutamyl- ϵ-lysine bridges which is Ca^{2+}-activated *trans*-glutaminase mediated. Such anchoring besides limiting protein motility may also cause deeper insertion within the bilayer and

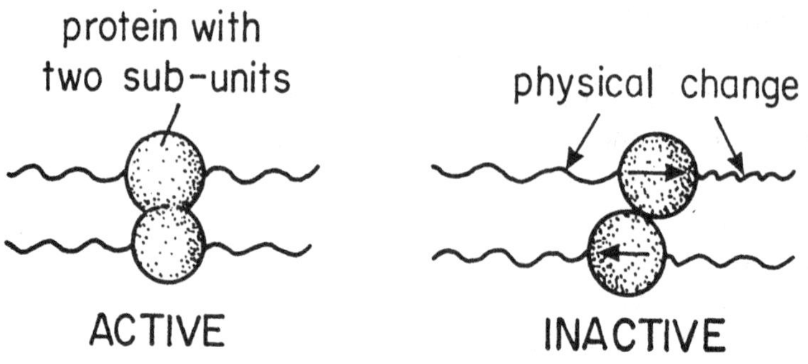

Fig. 5.5. Change of viscosity in one leaflet of the bilayer may cause a stereochemical change in one of an enzyme's subunits thus inhibiting its biological activity.

thus less exposure for interaction with effectors on the membrane's exterior.

The acidic nature of several hormones (IAA, GA and ABA) may induce displacement of bridging Ca^{2+}. Low hormone concentrations would decrease membrane rigidity and promote liquid crystalline configurations. At higher ones the Ca^{2+} stabilizing effect would be lost, anchored protein hormone receptors could be ejected from the bilayers and released Ca^{2+} may trigger the "phospholipase-lipoxygenase cascade" (Chapter 12) resulting in ethylene production and senesence. In all these instances, high concentrations of the divalent cations will cause an undesirable rigidification of the membrane, while under very low concentrations the membrane would experience destabilization stemming from a *supra* fluid state.

Another biophysical effect on integral protein occurs if only one leaflet of the bilayer undergoes a change of some physical parameter while the other remains unchanged. In such a situation a conformational perturbation may be experienced by the protein and resultingly the latter's biological activity may be impaired (Schroeder, 1985). This indeed may be the case in a membrane-contained enzyme comprised of two subunits, each being associated with a different leaflet subjected to unilateral change as diagramatically depicted in Fig. 5.5 and which may apply to adenyl-cyclase.

5.4. Malondealdehyde-mediated cross-linking of membrane proteins

Membrane-contained protein motility – and hence biological activity – may be curtailed by detrimental cross-linking with *malondealdehyde* which can be produced during the process of membrane lipid peroxi-

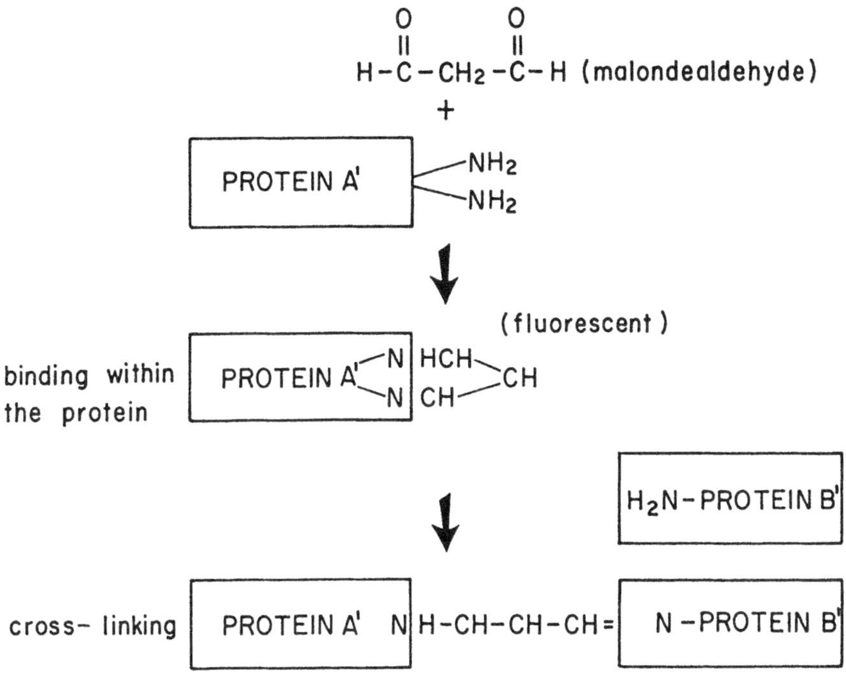

Fig. 5.6. Malonaldehyde-mediated cross-linking of proteins: malonaldehyde being a product of lipid peroxidation (after Kikugawa and Beppu, 1987). This mechanism was established in tea (*Thea bromia*) leaves. Cross binding of this nature immobilizes integral proteins and concomitantly inhibits their biological activity.

dation. As typically found in tea leaves, Fig. 5.6 indicates how this may be brought about. In the absence of *α-tocopherol* (Vitamin E) which serves as an antioxidant the essentially immobile protein:malondealdehyde complex is more readily formed (Manwaring and Csallany, 1988).

In various research laboratories it is a common practice to utilize specific enzymes' activities as biological markers in order to indicate purity and specificity of various membrane preparations by two-phase separation, sucrose gradients, ultracentrifugation, etc. Thus for example respiratory enzymes such as *cytochrome oxidase* are markers for mitochondrial membranes, *vanadate-sensitive ATPase* activity is associated with the plasmalemma, *nitrate* (*NO_3*)*-sensitive ATPase* typifies the tonoplast, and photosynthetic enzymes are markers for chloroplast enzymes (Oleski et al., 1987; Widell and Larsson, 1990). However in view of what has been described in the previous paragraph concerning fluidity-related physical changes, it is pointed out that lack of marker enzyme activity does not of necessity rule out the possibility of a pure membrane isolation, since extraction procedures may well involve enzyme inactivation. In such cases immunoassay or cytostaining, less commonly practiced, are preferable (Gahan, 1984, 1989; Schroeder, 1985). In both latter cases artefactal loss of biological activity of marker enzymes does not usually hinder antibody-antigen determinant recognition or a specific staining response.

5.5. Proteolipids

A further class of complex integral membrane proteins is the proteolipids which, as the name implies, are comprised of both proteins and lipids. This specific type of lipid is found in biological membranes especially in the brain where it constitutes approximately 20% of the total. Other tissues in which proteolipids are found are in membranes of nerve axons of the liver and of kidneys. In plants the presence of these proteins – which due to their lipid moieties are hydrophobic – has been reported in chloroplast membranes (Pick and Racker, 1979) and in endosperm tissue of wheat and barley (Shewry et al., 1984).

Biochemical analysis has revealed that their protein section is comprised of 276 amino acid residues about half of which are covalently bound to fatty acids. Moreover these proteolipids, like ATPase, traverse the membrane four times in a typical looping fashion. Regarding function of lipoproteins it appears that they are linked to activation of the Ca^{2+} pump – Ca^{2+}ATPase – and also of the proton pump – H^{+}ATPase. The involvement of proteolipids in plant membrane physiology has been only partly investigated and the putative analogies to mammalian proteolipid structure and function is at present not proven.

5.6. Ionic channels

It is now becoming increasingly apparent that permeability properties of biological membranes, such as ion selectivity, voltage dependent conductances (some of which have been dealt with earlier in this chapter), stereoisomer discrimination and active transport are not inherent properties of lipid bilayers, but rather of proteinaceous pathways inserted into and through the supporting structure. Following are descriptions of some of these pathways which have been studied in various categories of plant membranes.

Ionic channels, which essentially are integral membrane proteins, are proteins possessing constant thermodynamic motility and through

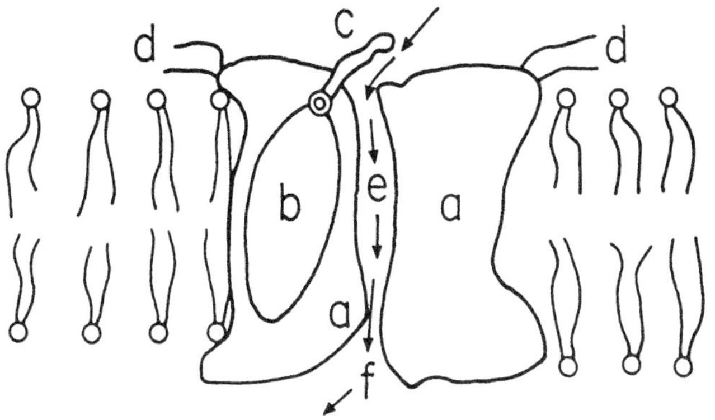

Fig. 5.7. A typical ionic channel. *a* channel protein, *b* sensor, *c* gate operated by voltage, hormones, light, stretch or specific ions, *d* anchoring protein, *e* hydrophilic channel, *f* selective filter (hypothetic) (after Satter and Moran, 1988).

which specific ions can penetrate. Ion movement through membranes and channeling physiology has been aided by the "*patch-clamping*" technique which was pioneered at the University of Göttingen by Neher and Sackman almost two decades ago. This technique is based on the ability to bring a small surface area of membrane protein (the "patch") in close contact with the tip of a finely extruded glass microelectrode (the "clamp") and to measure electrophysical changes in response to variation of ionic or microenvironmental changes of either the host cell or organelle or of the electrode-contained fluid. This technique has been applied to a wide variety of plant membranes allowing a highly sensitive recording of ionic contents both at the single channel level and from whole cells or organelles (Hedrich and Schroeder, 1989).

Ionic movement in these channels can broadly be categorized under four headings:

1. *Uniport* – transport of only one ion in one direction
2. *Symport** – transport of two ions in the same direction
3. *Antiport* – transport of two ions, each in an opposite direction
4. *Non-selective* cation channels activated by a specific cation or biophysical change.

The "open" situation of the channel (See Fig. 5.7) allows the specific ion to penetrate via a diffusion based mechanism which may be three to four times faster than carrier-mediated ion transport. The duration and frequency of opening depend upon stimuli which include electric charge, light, hormones, Ca^{2+} concentration, elastic stretch, etc. This opening and closing of channels has metaphorically been likened to a gate and hence the term *gating* which is invariably utilized

*A transmembrane glucose/H^+ symporter, pertinent to source-sink relationships and phloem unloading, is found in leaf mesophyll cells.

in channel terminology. Pursuing the metaphor further, the "gate-keeper" does not necessarily have to be the agent which directly passes through the channel but can be a secondary factor; e.g. in stomatal physiology K^+ channels are voltage dependant (Schroeder et al., 1987). In membrane of the moss *Funaria*, Saunders (1986) has reported an interesting dependence of Ca^{2+} channeling upon the phytohormone cytokinin.

Ionic channels have been found in most categories of plant membranes (Moran et al., 1984; Hedrich and Schroeder, 1989), but probably because of the medical and therapeutical interest in the mammalian and in particular human cell membranes (e.g. in the kidneys, nervous system, digestive tract, etc.) plant research has been comparatively less studied and less is known. The representation of a typical plant ion channel seen in Fig. 5.7 therefore remains hypothetical; nevertheless it has been quite clearly established that ion channels belonging to the four previously listed archetypes *are* present in higher plants, and that while they exhibit either anionic or cationic specificity, in certain instances may also be non-selective.

5.7. The possible structural and biophysical basis of ion channel specificity

Employing in vitro models, a rather interesting and novel approach which may contribute towards the understanding of ion channel specificity has been taken by Lear et al. (1988). Basing upon information drawn from in vivo experimentation with biological ionic channels these workers found indications suggesting that these channels are comprised of bundles of a given number of helical proteins possessing specific amino acid sequences. In a model they constructed (See Fig. 5.8) it was found that a peptide comprised of leucine and serine possessing a length of only 21 amino acids forms an ionic channel similar in properties and longevity to the acetylcholine receptor. It is noteworthy that a similar peptide comprised of only 14 amino acids apparently did not span the membrane bilayer and resultingly was devoid of channel activity. By manipulating the 21 amino-acid-containing peptide and changing the serine-leucine sequencing, a proton (H^+)-specific channel was synthesized. The structure of the individual segments was as follows:

$$(H_2N\text{—Leu—Ser—Ser—Leu—Leu—Ser—Leu})\text{—}CONH_2$$

These segments correspond to the cylinders seen in Fig. 5.8 and

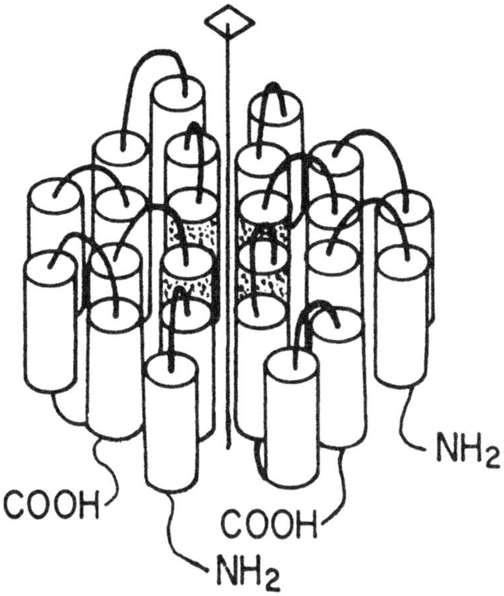

Fig. 5.8. A hypothetical biophysical structure of an ionic channel. The cylinders represent peptides comprised of 21 (3 × 7) leucine or serine sequences. The heavy lines are non-helical peptide segments (after Lear et al., 1988).

represent the helical regions. According to this model, the interior of the channel has a "lining" of 4–5 of such cylinders which span the bilayer from side to side. In this construction the *hydrophilic polar amine groups* face inwards and they are believed to impose specificity to the channel. Besides the above possibility comprising of 4–5 leucine-serine subunits, according to Lear and co-workers, another exists whereby the channel interior may be lined not with polar residues, but rather with *amphiphilic residues* comprised of different amino acid residues – these being serine, threonine and cysteine. By varying amino acid sequences and by interplay of different amino acids which are aligned, along the channel's interior it is suggested that ion specificity can be obtained.

The above concept providing a structural/biophysical explanation, albeit theoretical is of considerable interest since it provides a firmer basis to the basic understanding of ion specificity than the ephemeral "gating" concept which is no less theoretical. Further research will doubtless shed more light on the degree of pertinence of this approach to channeling in both plant and mammalian membranes. We point out that the detailed research carried out by various research laboratories, the world over, on the *acetylcholine receptor-operated* Na^+ *channel* (Fig. 5.9) in nerve cells suggests that Lear's theoretical

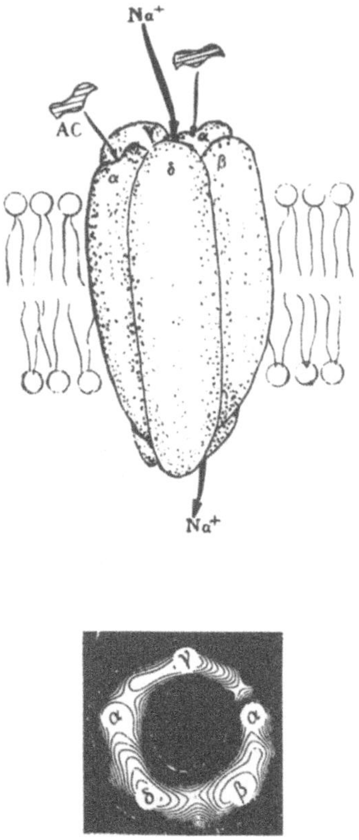

Fig. 5.9. The acetylcholine (AC) receptor and the Na^+ channel. *Above:* Receptor comprised of several protein subunits, each designated by a different Greek letter. The α-subunits serve as the acetylcholine receptor site; upon binding of the AC to these subunits, a Na^+ specific ion channel is opened. *Below:* Top view of the channel as obtained by X-ray crystallography (after Popot, 1987).

model may not be far removed from an in vivo situation. In keeping with this speculation, Mellor and Sanson (1990) have recently reported that *mastoparan*, a wasp venom constituent, as well as a synthetic venom analogue, cause the formation of two protein helices in a membrane, these eliciting specific channeling properties to K^+; moreover these helices bind to calmodulin and also activate G proteins. They also found that the mastoparan venom constituent contains *14* residues of amino acids.

Ion specificity, according to the above approach, lies in a specific biophysical and steric configuration of the channel (or its gate) lining, which is a function of amino acid sequencing which in turn allows certain ions to pass through while obstructing entry of others whose steric configurations exclude passage through the channel.

Since the overall aim of this text is to provide a basis for basic biophysical underlying membrane structure and function – and in this section, of ionic channels – I do not aspire to provide the reader with a detailed list of specific channels and plant species in which they have been found, but rather to highlight possible common denominatory mechanisms. However to provide the interested reader with leads for further pursuit of the subject, several salient and typical examples are mentioned herewith:

K^+ channels: K^+ channels in plants were originally detected in stomatal guard cell protoplasts and subsequently in several other membranes including thylakoids. A common factor apparently is regulation by membrane potential. As well known to plant physiologists K^+ fluxes are of major importance in stomatal movements, where K^+ channel-associated transport fully accounts for this ion's fluxes (Hedrich et al., 1990). A similar situation exists in leaf pulvinus movement of the "touch-me-not" (*Mimosa pudica* and *Albizzia* spp.) as indicated by the pioneering work of the late Ruth Satter and colleagues. A somewhat different K^+ channel gated by Ca^{2+} and not membrane potential has recently been reported in *Mougeotia* membranes (Low et al., 1990). These are but three examples of many.

Ca^{2+} channels: It has been shown that the development of lily pollen tubes is dependent upon free passage of Ca^{2+} through specific Ca^{2+} channels which can be blocked by the Ca^{2+} channel blocker *nifendipine*. These channels contribute to the maintenance of a polar gradient of intracellular Ca^{2+} and it has been proposed that the opening of these channels relies upon voltage-dependent gating very much as in animals (Reiss and Herth, 1985). Specific Ca^{2+} channels of a different type (detailed later under "stretch operated channels") have been reported in stomatal guard cells (Cosgrove and Hedrich, 1990). An interesting case of *Ca^{2+}-activated non-selective cation channels* has been found in endosperm membranes of *Haemanthus* and *Clivia* (Skoeckel and Takeda, 1990) – it is noteworthy that these channels are reported to be impermeable to anions.

Na^+ channels: The vacuole is of pivotal importance in regulation of cellular osmotic pressure potentials, as well as in sequestering potentially harmful ions such as Na^+, and Cl^- under salt stress. It thus is to be expected that the tonoplast would contain channels for these ions, and indeed in tonoplast membranes of *Plantago*, Maasthuis and Prins (1990) have documented the presence of a specific Na^+ channel which is gated by both Ca^{2+} and voltage. In terms of salt tolerance it is of interest that under conditions of senescence and of NaCl stress, opening of these channels decreased. It would be of interest to ascer-

tain the degree of similarity or dissimilarity of this plant Na^{2+} channel to the acetylcholine channel in mammalian nerve tissue as depicted in Fig. 5.9.

In halophytic plants an adaptation mechanism is to remove surplus Na^+ and also sometimes K^+ ions from the cytosol into the vacuole via the tonoplast. This detoxifying mechanism has been studied by Leonora Reinhold's laboratory at the Hebrew University in Jerusalem (cf. Hassidim et al., 1990) in the extreme halophyte *Atriplex halimus* which is a predominant species of the Jordan valley and Dead Sea littoral salt marsh vegetation in Israel. Research conducted on this species revealed Na^+/H and K^+/H^+ antiporters in membrane vesicles isolated both from the tonoplast and from the plasmalemma. The plasmalemmal antiporter functions to extrude the cations to the leaf's exterior upon which salt accumulates and thus, in addition to vacuole accumulation, serves as a further salt-coping strategy. Similar tonoplast Na^+/H^+ and K^+/H^+ antiporter ion channels have been found in cell suspensions of sugar beet which belongs to the same family – *Chenopodiaceae* – as the *Atriplex* and which is also halophytic to a certain degree (Pantoja et al., 1990) as well as in cotton which while not a halophyte nevertheless manifests ability to withstand mild salt stress.

The above paragraphs deal with *cation* channels, however plant research has also revealed the existence of membrane-located *anion* channels in particular *Cl^- channels*. Such channels have been detected not only as may be expected, in marine algae such as *Chara* (Coleman, 1986) but also in higher plants such as *Samanea saman* and *Asclepias tuberosa* tissues (Schauf and Wilson, 1987; Moran, 1990). These channels presumably play a role in osmo-regulation which is essential for maintenance of optimal growth conditions in these species.

Stretch operated channels: Perhaps the most interesting recent development in plant channeling mechanisms is the demonstration of stretch operated channels in cells which are prone to piezobiophysical changes. As most pertinently stated by Morris (1990) in her topical review of the situation in mammalian and lower Animal Kingdom membranes:

> cells (like cats) are seldom static. They grow, migrate, contract, change volume and shape, reorganize internally, divide, and are subject to all manner of compression, shearing and stretch. Evidence is growing that most cells have ion channels potentially capable of monitoring and regulating active and passive variations in cellular mechanics. The ion channels in question are mechanosensitive, that is, their open probability depends on stress at the mem-

brane. Channels of this type were postulated decades ago as a way to account for mechanoelectrical transduction in muscle spindles, and in crustacean stretch-receptors, Pacinian corpuscles and other specialized mechanoreceptors. What was not anticipated was the picture now emerging, with mechanosensitive channels present as membrane components in cells, rather than being confined to exotic mechanoreceptor specialist ones.

Mechanosensitive membrane channels of various ion selectivities have been found in a multitude of animal cells of vertebrate and invertebrate origin as well as in bacteria and fungi. Presently these channels are classified, in what may be merely a phenomenological manner, under two overall categories – *stretch activated channels* and *stretch inactivated channels*.

In plants a stretch activated anion channel was found to be present in tobacco protoplasts (Folke et al., 1988). Two other typical cases are mentioned, the first being in broad bean (*Vicia faba*) guard cells where, as in like stomatal complexes, guard cells undergo diurnal cyclic changes of turgor. Cosgrove and Hedrich (1990) report that upon being exposed to "stretch", a *non-specific activation* may occur of cation, anion and Ca^{2+} channels. These workers propose that these channels may contribute to turgor regulation directly by ion influx, and indirectly by activation of non-stretch channels through regulation of cytoplasmic Ca^{2+} and voltage. The second example, as reported by Moran (1990), is the stretch activated channel in the plasmalemma of *Samanea saman* pulvinar motor cells. Moran has found that this membrane houses a stretch-associated K^+ selective channel which opens in the swollen cell and which in turn allows passive efflux of Cl^-, thus causing change of components of the cell's osmoticum.

Biophysical aspects of stretch gating: The biophysical principles connected with osmosis/turgor-regulated channel gating have been discussed in detail by Finkelstein (1987). He has pointed out that osmotic regulation of channels will occur if there are volume changes associated with the transition of a channel from the open to the closed state. The possible magnitude of this effect may be learned by envisioning a water-filled cylindrical channel of radius r and length L, and assuming that in going from the open conducting state to the closed non-conducting state, the channel lumen is obliterated throughout its length. In this transition a volume ($V = \pi r^2 L$) of solution is expelled from the channel into the surrounding medium, and conversely this volume enters the channel from the surrounding medium when the channel undergoes the transition from the closed to open state. There is thus a $V\Delta P$ term associated with the free energy change of the transition between the open and closed state of the channel. We can write f for the ratio of time spent in the open state to time spent in the closed state (which, for an ensemble of channels is the ratio of the number of channels in the open state to the number in the closed state).

Employing estimated dimensions of a typical biological channel (r = 4Å and L = 30Å) Finkelstein has calculated that if the surrounding medium does not contain an impermeant solute the pressure term makes no contribution to f. However f is decreased about 4-fold when the osmolarity of the surrounding medium is raised to 1 molar with an impermeant solute; that is, the ratio of the time spent by the channel in the open state to that spent in the closed state *is reduced by a factor of 4.*

He has moreover indicated that the effect of *osmotic pressure* on channel gating described above is very different from possible effects of *hydrostatic pressure* on gating. The former depends on volume changes in pore content in transition from the open to closed state and in the *difference* in hydrostatic pressure between water within the pore and that in the surrounding medium. Externally applied hydrostatic pressure is transmitted equally to water within and outside the pore, and therefore does not alter the contribution made by volume changes in pore content to the free energy change of the transition between open and closed state. Any effects of externally applied hydrostatic pressure on the free energy of this transition will be associated with compressibility of, and small partial molar volume changes in, the protein (and lipid) forming the channel. Much larger hydrostatic pressures are required to produce measurable effects on channel gating than the osmotic pressures considered above.

While the overall concept of stretch operated channels and their function in osmo-regulation is a tempting and elegant way to explain certain aspects of plant turgor change, it should be remembered that the above and other reports are primarily based upon experimentation where "stretch" is applied on patch-pipette experimentation employing the above-described technique. These measured effects are doubtless real and well founded, but before unequivocally stating that localized pressure applied during patch clamping reflects an in vivo biological situation, more evidence will have to be presented. This is all the more so since in their natural state biological membranes subjected to extreme turgor or pressure changes often resort to mechanical stretch-resistance strategies. These may include rigid cell walls in the case of stomatal plasma membranes, which act on the membrane's exterior, cytoskeletal elements acting on the membrane interior or even of the limited presence of a membrane-contained rigidifying phospholipid or sterol. Thus in addition to the patch clamping technique, in solving this problem others too will have to be applied or developed, since patch clamping by its very nature is generally applied to cell wall-deshabilled protoplasts.

5.8. Contentions against ion-channeling and pumping

With the object of providing an up-to-date objective overview of the "state of the art" in biological transfer of ions and ion pumping, the author deems it incumbent upon him to present a viewpoint which, while not based on experimentation on plant tissues, is of general physiological interest and may be considered by plant scientists. While presenting this view we stress that most, but not all, scientists have reservations about the theory to be described:

Ling (1988, 1991) formulated a "*polarized multilayer theory of cell water*" in which the physiological importance of ion pumping is contended. The work of Ling and colleagues, conducted over a period of 40 years, has encompassed virtually all of the physiologically important cations and anions as well as other metabolites.

This theory envisages several layers (up to 10 – based on theoretical calculations) of water molecules arrayed along proteins in their extended state, the water in this particular state comprising ca 95% of the total. Small ions such as Na^+ and K^+ and molecules such as urea can fit into the dynamic water structure of the polarized water and therefore may *freely* pass along the protein:water assembly. Only large molecules that cannot fit this water structure are excluded. Thus a "size rule" is obtained – *hydrated* Na^+, sugars and free amino acids which as a rule are excluded from the water of living cells are large and do not fit the polarized water structure and are therefore partially excluded. Ling (1988) has moreover claimed that in muscle tissue, were ATP energy-linked pumping employed to account for observed Na^+ changes, the minimally needed energy is 15 to 30 times more than the maximally available energy. On the basis of these and other experiments Ling claims that certain generally accepted basic tenets of ATPase-driven ion-channel pumping are untenable.

5.9. Coated pits and coated vesicles

Membrane turnover, as envisaged by Steer (1988), employs vesiculation as a major, if not a sole mechanism and in his opinion occurs as a consequence of secretion which incorporates new membrane

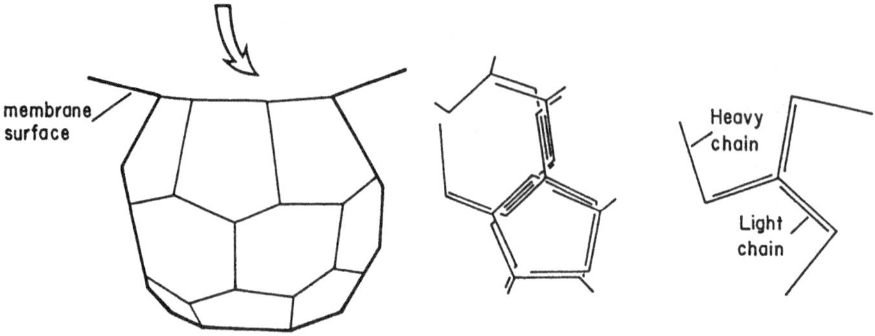

Fig. 5.10. Structure of a coated pit. *Left:* the "triskelion" comprised of light- and heavy chained polypeptide sub-units. *Center:* assembly of triskelions into hexagons and pentagons. *Right:* polygonal structure of the coated pit surface (after Robinson and Hillmer, 1990).

into the cell surface, and of endocytosis which internalizes surface membrane. This interchange is purportedly mediated by *coated pits* (not to be confused with xylem pits which are structural elements in the plant's vascular system) and *coated vesicles*. These are localized membrane areas with polygonal assemblies of *clathrin*, a hexamer composed of 180 KD polypeptide subunits in an especial geometrial configuration known as "*triskelion*" (Fig. 5.10). Such coated units do not cover extensive portions of a membrane and are primarily confined to regions where vesiculation is occurring or is incipient. Emons and Traas (1986) have estimated that higher plant cells have on the average 5,000–10,000 coated pits per cell. The role of coated pits and vesicles in the regulation of endocytosis and membrane traffic in plant cells is now well documented. The further role of uptake of specific molecules as occurring beyond doubt in animal membranes and which has led to the plasma-membrane located "*receptosome*" concept in animal physiology, has only recently found experimental support in plants in an as yet limited number of molecular species, these including polygalacturonic acid and oligo- and polysaccharide elicitors of plant defense responses to fungal infection (Horn et al., 1989), and may be involved in the hypersensitive response.

5.10. Plant hormone receptors

Hormone/receptor interaction constitutes the initial step in a series of events by which many hormones regulate cellular metabolism. Binding is not a priori evidence that the moiety which recognizes the hormones mediates hormone action. However, in certain cases a receptor role has been demonstrated which is directly associated with hormone action. The magnitude of hormone/receptor effect is subject to regulation by rate of receptor production, loss or relocation in membrane domains. Thus in a given situation high endogenous hormone concentrations associated with low receptor availability, or vice versa, are not conducive to optimal physiological response.

The present text does not aspire to review the above topic in its entirety, since various aspects of plant hormone receptor function extend well beyond the confines of the plant's membrane(s): it has quite clearly been established that certain proteins *not* located in membranes may answer to this overall definition. We shall therefore limit the present discussion to membrane-associated receptors and briefly mention other types where necessary. Those interested in a bird's-eye view of the plant hormone receptor field in overall terms are referred to the volume edited by Klämbt (1987) and the reviews of Guern (1987) and of Napier and Venis (1990).

Receptor proteins for all five major classes of phytohormones – auxins, gibberellins, cytokinins, abscisic acid and ethylene – have been discovered and it appears that the majority of these receptors are membrane-associated. Plant research in this field is still far-removed from the sophistication of mammalian receptor research as exemplified by the detailed structural clarification of the acetylcholine receptor in nerve synapses or of the prolactin (the milk promoting hormone in mammary glands) receptor. Moreover, despite the tempting surmise that this may apply to phytohormones as well as to mammalian hormones, no conclusive evidence has been presented that *endocytosis* based upon coated pit and vesicle shuttling – the "receptosome" mechanism in animal endocrinology – operates in plant membrane hormone recognition.

One basic conundrum in plant receptor research is that of the hormonal and physiological significance of the binding to specific protein receptors – or in other words, if the physiological response attributed to a given hormone is indeed the outcome of binding to the said acceptor. The possibility thus arises that the two processes viz. binding to protein, and physiological effect, are not of necessity linked and that binding may only be a structural phenomenon. Moreover as shown by Shimomura et al. (1986) the affinity of the auxin-receptor isolated from maize shoot membrane for auxin is *200 times more* than that required for optimal shoot elongation.

A second query is to the necessary pertinence – following the pattern in animal endocrinology – of a quest for *protein* receptors –

and in addition as a rider to hormone receptor binding hypothesis, why does a biological and not necessarily a physical explanation have to be offered. In the past Jones and Paleg (1984) have presented in vitro evidence that auxin binding could be via attachment to phosphatidylcholine. Moreover both in vivo and in vitro experimentation in the author's laboratory (Landau and Leshem, 1988; Leshem et al., 1990) has indicated that both auxin's and abscisic acid's physiological action may include biophysical aspects; these in no way being connected with proteinaceous receptors. These reports however are not necessarily exclusive of the protein-receptor approach especially since current opinion is that phytohormones have more than one mode of action and possibly possess multiple binding sites.

To return to the physiological significance of hormone:protein binding, the author endorses the approach taken by Guern (1987) that a clear distinction should be made between a *hormone receptor* and a *hormone binding protein*. The significance of the hormone:receptor binding required for physiological response is important in understanding why high endogenous levels of hormone alone are not sufficient to elicit responses if these do not go hand-in-hand with parallel levels of receptor.

In following sections I shall address some of the above questions and shall endeavor to discuss certain pertinent aspects of hormone receptors of the five major aforementioned classes. In keeping with the topic of this chapter – "Membrane Proteins" – the discussion will be limited to proteinaceous hormone receptors while the phenomena of other types of hormone reception and recognition will be dealt with elsewhere in this text.

5.10.1. *The auxin receptors*

Research carried out in various laboratories the world over has shown that there are at least *three* auxin receptors each of which apparently is associated with a different auxin-evoked effect, this hormone being well documented to elicit pleiotropic responses of plant growth and differentiation. The basic research conducted at the University of Leiden in the Netherlands (Mennes et al., 1990) typically represents the situation. According to this group the auxin receptors are of three types:

a. *A membrane located receptor with a marked affinity for auxin* appears to be identical or similar to the one isolated by Lobler and

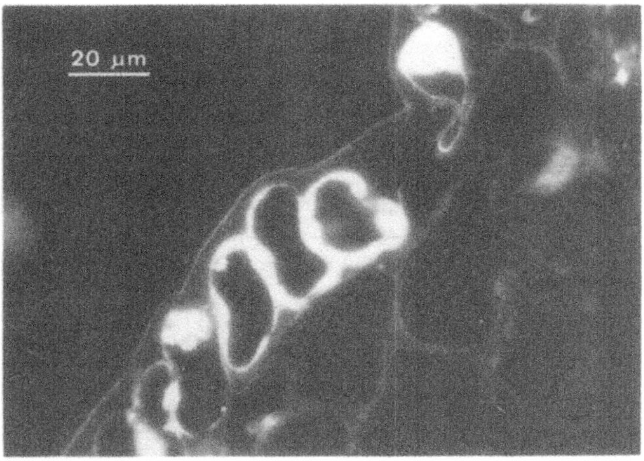

Fig. 5.11. Plasmalemmal location of the auxin receptor protein antibodies in corn shoot cells. The antibodies are seen as fluorescing regions (after Lobler and Klämbt, 1985).

Klämbt (1985) of Bonn. This receptor is a peripheral protein located on the plasmalemmal bilayer's outer leaflet although it is initially formed in the Golgi apparatus from which it is transported to the plasmalemma (see Fig. 5.11). The receptor is an example of one of the very few cases where it has unequivocally been demonstrated that receptor protein:hormone binding is of physiological significance, this being expressed in inducing curvature of split cereal coleoptiles in a classic bioassay (see Fig. 5.12). Em-

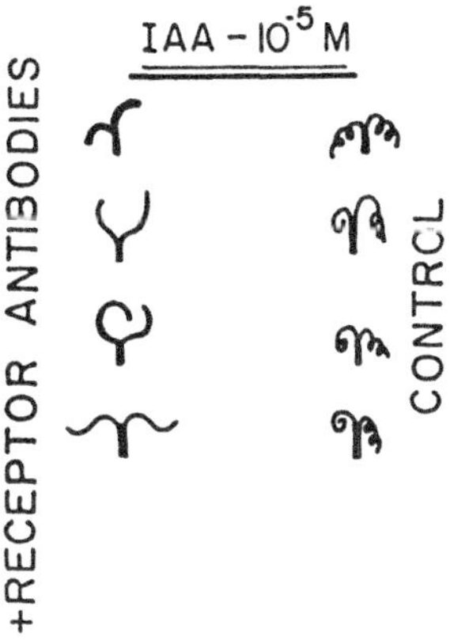

Fig. 5.12. Evidence that auxin binding to a receptor protein induces a physiological response. Longitudinally split corn coleoptiles curve when treated with auxin (right column). When auxin receptor antibodies were added, degree of curvature markedly decreased (left column). After Lobler and Klämbt, 1985.

ploying immunological techniques, antibodies to this particular receptor protein were prepared and, when applied to corn shoot tissue, inhibited a typical auxin (indoleacetic acid – IAA) response. It is of interest that this receptor also reacts with two other compounds – naphtylphtalamic acid (NPA) and tri-iodobenzoic acid (TIBA) both known to inhibit auxin's effects by blocking the hormone's transport and other physiological responses.

It is moreover possible that this receptor has much in common with that found in corn coleoptiles by Venis's group in Britain (Venis et al., 1990). As in the above case this receptor proved to be a peripheral protein: it is a dimer with 22 kD subunits and also includes a glycosyl moiety. Monoclonal antibodies to this protein indicated its presence not only in corn but also in other monocots, as well as in dicots. Venis also reports that the amino acid sequence of the receptor includes several hydrophobic species which aid its anchoring in the membrane. Physiological importance of the hormone/protein binding was noted by the identification of a similar receptor located in the plasmalemma of tobacco mesophyll cells which upon binding with auxin leads to activation of H^+ ATPase (Barbier-Brygoo et al., 1989). A 43 kD glycoprotein thought to be a dimer of the above-mentioned 22 kD subunit has now been identified as a strong candidate for the auxin receptor in etiolated maize shoots (Jones, 1990).

b. *A membrane protein associated with the rooting response.* The affinity of this receptor for auxin is markedly less and hence a possible explanation why relatively high auxin concentrations, considerably higher than endogenous levels, are required for evoking horticulturally important rooting responses in cuttings.

c. *A soluble cytoplasmic non-membrane contained receptor.* This receptor is claimed to convey a message to the genetic apparatus and to promote transcription of a particular type of mRNA which in the final outcome results in the typical promotion of cell division elicited by auxin application (Mennes et al., 1990).

In agricultural practice the use of selective herbicides in cereals to kill off all non-cereal ("broad-leaf") weed species is of widespread practice. These selective herbicides are auxin-based (NAA, IAA, and MCPA) and at a given concentration range cereal fields – wheat, barley, rice, etc. – manifest selective resistance to herbicide application. Amongst other mechanisms contributing to this inherent resistance, the presence, or conversely the absence of auxin receptor proteins has been mooted, but so far this interesting possibility remains hypothetical.

5.10.2. *Cytokinin receptors*

Compared to what is known about auxin receptors, the information regarding cytokinin receptors is sparse. Biochemical characterization together with genetic studies has revealed a cytokinin binding protein in cereal seed tissue – the scutellum and the coleorhiza being the sites of occurrence (Brinegar et al., 1988). This receptor peptide comprises ca 9% of total protein in the seed embryo and contains three identical sub-units each having a M.W. of 54,000. Limited binding of cytokinin and of the synthetic cytokinin benzyladenine, has been found in peas, soya, and cucumber cotyledons (Jayabaskaran, 1990) but no physiological effects were reported to be associated with the binding in the above cases.

Recently some Chinese workers (Huang et al., 1990) have isolated a cytokinin binding protein from the chloroplast outer membrane. They reported that the addition of benzyladenine and its binding to the chloroplast membrane receptor causes energization of this organelle apparently via promotion of phosphorylation responses linked to delayed reflectance of photosynthetic light.

Another cytokinin receptor, *not* located in the membrane, but with demonstrated physiological pertinence has been found in barley foliage (Romanov et al., 1988).

5.10.3. *Ethylene receptors*

As well reviewed by Sisler and Goren (1981) it has been clearly demonstrated that ethylene binds to a protein receptor and that such binding is a prerequisite for the manifestation of hormonal effects. Subsequent experimentation has shown the presence of *three* receptors, varying in temporal action (Sisler 1990). Sisler clearly distinguishes between the possibility or site of ethylene binding which is of biological importance, and the other(s) which demonstrate binding of as yet an unknown nature. It is currently not known which of the three receptors is the physiologically active one. In the context of ethylene: receptor physiology it is noteworthy that the unsaturated hydrocarbon gas *norbonadiene* which acts as an ethylene inhibitor, also attaches to cytokinin receptors, and thus possibly acts in a competitive manner with ethylene to forestall the latter's effects.

The total number of ethylene binding sites gradually increases with time and peaks when tissue is still relatively young and afterwards

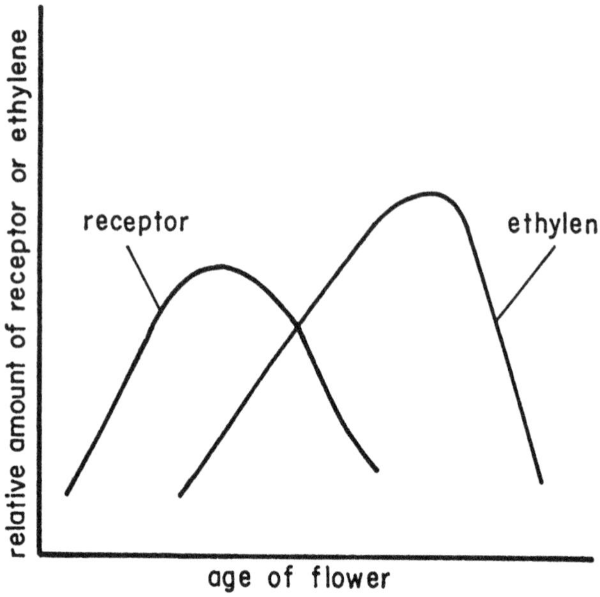

Fig. 5.13. Kinetics of appearance of ethylene binding sites and ethylene production in microsomal membranes of carnation petals (after Thompson, 1983).

declines, whereas maximal ethylene production is considerably later. Research carried out in John Thompson's group in Waterloo, Canada has indicated that in cut carnation flowers the number of ethylene receptor sites is highest when the flower is still closed or undeveloped, and that only later, when the number of receptor sites declines, does ethylene increase (see Fig. 5.13). This binding provides an explanation of the marked sensitivity of young carnations to contact with exogenous ethylene with which they may come in contact during transportation and marketing.

It would be interesting to ascertain whether the endogenous ethylene inhibitor *feruloylglucose*, isolated and elegantly characterized in Dumbroff's laboratory (cf. Shih et al., 1989) from microsomal membranes, also acts by competitive binding to an ethylene receptor as does norbonadiene. That this may be so is suggested by noticing that quantification of the feruloylglucose indicated a similar temporal pattern to that of receptor sites as seen in Fig. 5.13 – amounts being maximal at the preclimacteric stage of carnation petal development but falling sharply upon onset of climacteric and senescence.

5.10.4. *The gibberellin receptor*

As in the case of the other phytohormones, gibberellins also appear to have receptors, but as presently known this receptor category is

not located in the membrane but rather in the cytosol in which both soluble target and non-target proteins appear (Srivastava, 1987). As in several of the above-mentioned instances, evidence presented indicates that while a specific binding protein doubtless exists, its hormonal pertinence still awaits demonstration.

5.10.5. *The abscisic acid receptor*

Proteins in maize guards cell plasmalemmas have been reported by Hornberg and Weiler (1984) to serve as high affinity sites for ABA binding. Three different polypeptides could be identified as being involved in the binding by employing light activated crosslinking of ABA to the binding sites. Also in maize, McCarty et al. (1989) have reported a wild type allele which codes for an ABA receptor and which may be absent in mutants. However the possibility that ABA may not only bind with membrane protein sites but also with *lyso*-PLs or disaturated PL species in membranes is broached in Chapter 9.

5.11. Lipid transfer proteins

As shown in Chapters 2 and 3, the glycerolipid and sulpholipid membrane components do not assemble at random and a certain pattern is followed. This pattern is often symmetrical, and within an individual monolayer is of a given biochemical constituency. In terms of membrane biogenesis the question is how this non-random assembly takes place. A clearcut answer is still wanting – one as yet unproven hypothesis envisages a "template-directed" assembly. Moreover in normally developing cells it is known that a membrane phospholipid located on one leaflet has virtually no chance of flip-flopping to the other. Notwithstanding these limitations it has been demonstrated that interchange of other plant membrane components may be mediated by a process designated *intermembrane exchange*.

Pioneering work initiated at the Pierre and Marie Curie University in Paris conducted by Kader's group and followed up in several other laboratories has conclusively established both the function, and to a certain degree, the structure of such exchange proteins. At the outset we state that these proteins are not membrane-located and thus may not be considered pertinent to the present text. I nevertheless have included this section since an overview of this group of proteins is necessary for an overall understanding of membrane architecture and function. The following paragraphs highlight major aspects of the interchange mechanism.

The fact that a phospholipid like PC is present in all plant organelles whereas the enzymes necessary for its synthesis are exclusively located in endoplasmic reticulum originally raised the idea that intracellular movements of phospholipids can occur. Movements of long-chain fatty acids have been studied in vivo between various membranes (Moreau et al., 1988). Since spontaneous movements of phospholipids which are weakly soluble in aqueous phases are unlikely, the other mechanisms could imply either transfer of vesicles or protein-carrier mechanisms. The discovery of soluble proteins, able to facilitate intermembrane movements of phospholipids in vitro, reinforced the latter hypothesis. These proteins, called *lipid-transfer proteins* (LTP), have been purified to homogeneity from various plant tissues such as spinach leaves, castor bean seedlings, maize seedlings (cf. minireview by Kader and Laurent, 1990). Their primary structure has been established and it has been found that these basic, 9–kD proteins (91 to 93 residues) have eight cysteines at conserved positions. A full-length cDNA has been isolated and characterized (Tchang et al., 1988). From this cDNA the presence of a "signal peptide" of 27 residues at the N-terminal end was deduced. A similar result was found with spinach, *Arabidopsis* and barley protein, and was recently identified as a LTP. These LTPs are active with various membranes, including mitochondria, plasmalemma or chloroplasts. This could explain the origin of phospholipids which are not formed in these organelles. It is especially the case with PC which is located in the outer leaflet of the outer membrane of chloroplasts.

LTPs can also "remodel" phospholipids e.g. they may remove an 18:1-containing PC from the outer leaflet of the outer membrane of chloroplasts, transfer it to the ER where it is reprocessed to an 18:2-PC and thereafter shuttle it back to the chloroplast. A similar shuttling phenomenon is also known to occur in potato mitochondrial microsomal* membranes. Arondel et al. (1989) stress that these LTPs are non-specific as to the phospholipid species they transfer and are moreover capable of transferring diacyl glycerols and even fatty acids (Rickers et al., 1985) and not necessarily intact phospholipids. Moreover in oil seeds – sunflower and rapeseed – it has been demonstrated that LTPs are also able to bind with oleoyl CoA, this suggesting a bifunctional

**Microsomes* – a non-specific term employed to describe membranes which sediment at *g* forces exceeding 100,000. It is estimated that a large proportion of the microsomal component is plasmalemma.

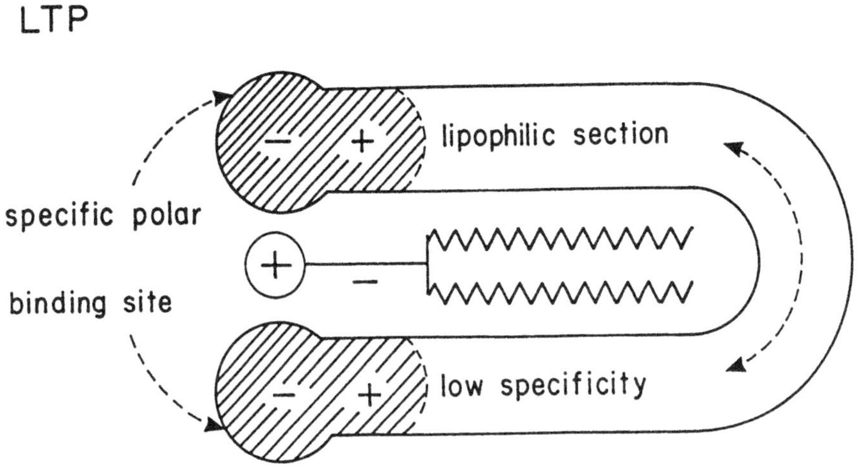

Fig. 5.14. The suggested structure of a lipid transfer protein (LTP). After Kader et al., 1990.

role in regulation of the intracellular pool of acyl CoA within plant cells (Kader et al., 1991). A formidable step forward towards the elucidating of the function of this group of proteins is the isolation of four isoforms, A to D, of nonspecific LTPs in castor bean seedlings, these isoforms being specific for specific seed structures: A and C in cotyledons, B in the endosperm and D in the seed axis and radicle (Yamada et al., 1991).

Fig. 5.14 represents a model that has been suggested for the stereochemical structure of LTPs. It is seen as a U-shaped protein harboring a "crevice" into which the fatty-acyl chains of the glycerolipid may be inserted. At the mouth of crevice, comprised of the two distal ends of the bent protein, there is a hydrophilic (lipophobic) polar site sterically and electrochemically favorable to binding of the headgroup moieties of phospholipids. This protein may thus serve as a shuttle service between various membranes within the cell and be of no small importance in the retailoring of membranes when the plant is subjected to stress, as well as in initial membrane biogenesis. This mechanism also affords an explanation why, despite lipid transfer, no flip-flop occurs – the shuttle operating through the cytosol and not across the bilayer.

A biotechnological direction in LTP research is to produce plant cultivars possessing high endogenous levels of these proteins, the rationale being that such LTP-enriched cultivars will be better equipped to withstand stress and to senesce at a slower rate. By using monoclonal antibodies to LTP it has been demonstrated that the LTPs present in widely divergent botanical species such as spinach, maize and cauliflower, are closely related (Grosbois et al., 1987).

With all the current interest, enthusiasm engendered and promise of LTP research, in all fairness it must be pointed out that some plant membranes are apparently *not* associated with such lipid transfer mechanisms – Schwitzguebel and Siegenthaler (1985) having documented their absence in certain cases. Whether LTP absence is plant species-specific or plant membrane category-specific is presently not known.

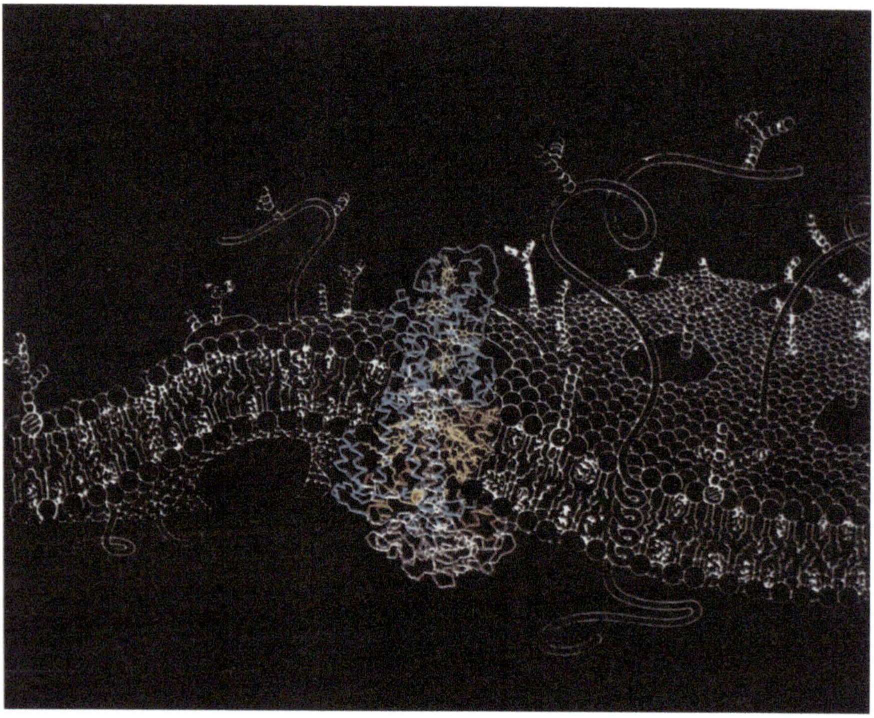

Colorplate Ia.

5.12. The photosynthetic reaction center

The early events of photosynthesis are mediated by a protein:pigment complex located in thylakoid membranes. Solar energy is trapped by light harvesting complexes and subsequently transferred to reaction centers. In the latter, with an efficient quantum expenditure, the trapped energy is harnessed for electron transmembrane transport,

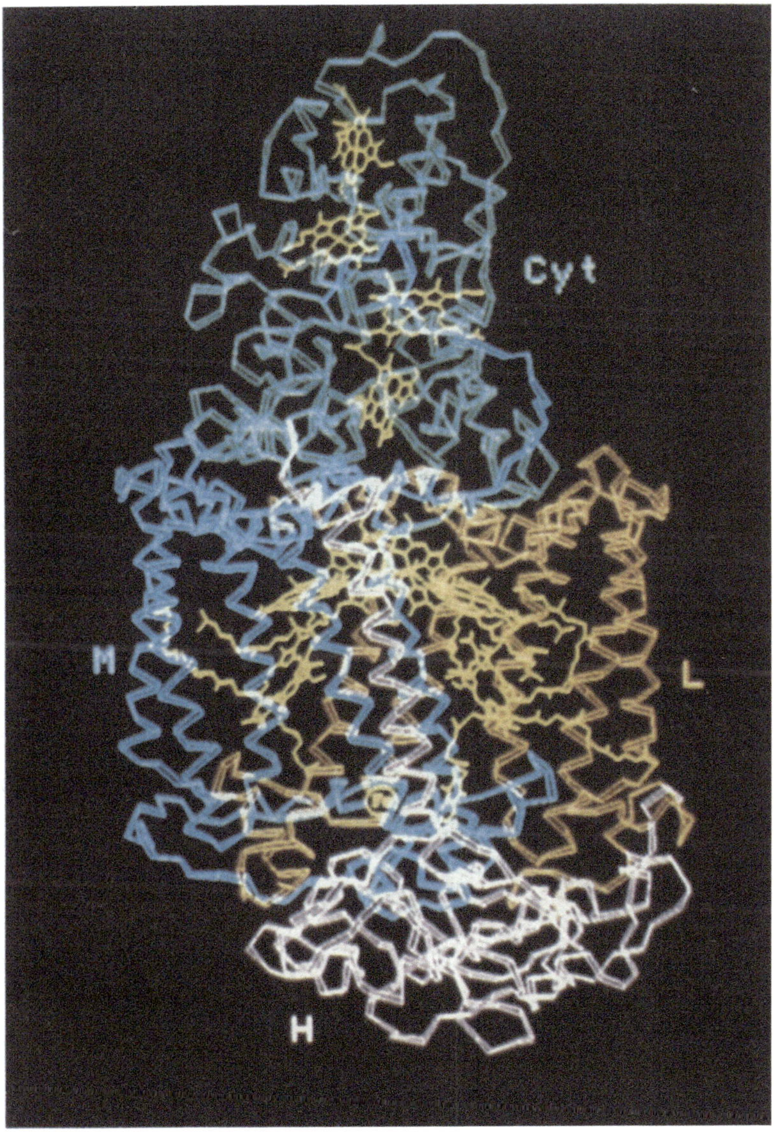

Colorplate Ib. The photosynthetic reaction center of the photosynthetic bacterium *Rhodopseudomonas viridis*. In the thylakoid membranes of higher plants similar reaction centers are believed to be found. Research enabling construction of this model led to the granting of the 1988 Nobel Prize. (This photograph is reprinted with the kind permission of Prof. H. Michel, Biological Unit, Max Planck Institute, Frankfurt, Germany). Color key: *Blue, brown and violet* – each color respectively presents the M, L and H protein subunits; *Green* – cytochrome subunits; *Yellow* – heme groups, photosynthetic pigments, quinone and non-heme contained Fe^{2+}.

and it is this step that may be considered as the initial one in the biosynthesis of photosynthetic assimilates. Plant senescence is moreover associated with a gradual decline in the ability of electrons to traverse the thylakoid membrane – a process attributed to age-related changes in photosynthetic proteins. In chloroplast senescence, protein denaturation in bean cotyledon thylakoid membranes causes decline of non-cyclic electron transport, while no changes in fatty acid composition, ratio of free to esterified fatty acids or in fluidity were encoun-

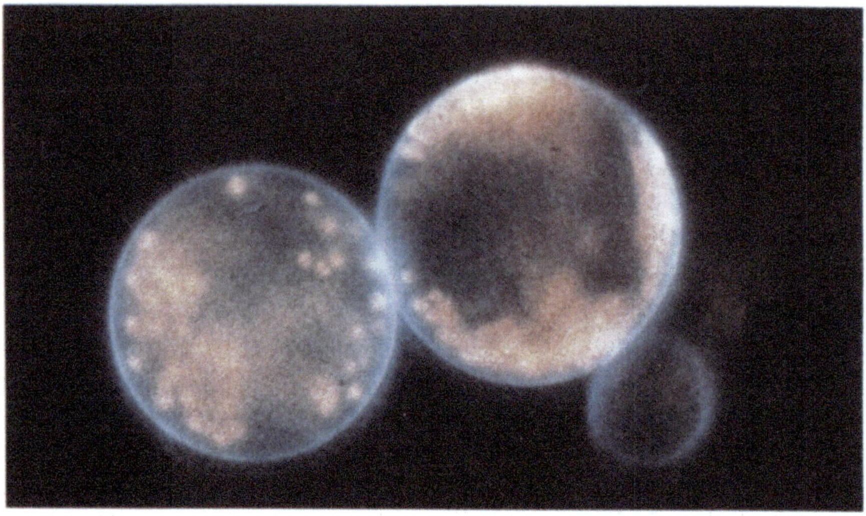

Colorplate II. TMA-diphenylhexatriene fluorescent probing of plasma membranes of rye protoplasts. (Photograph kindly made available by Dr. R. Langis, Department of Botany, McGill University, Quebec, Canada (see pg. 129).

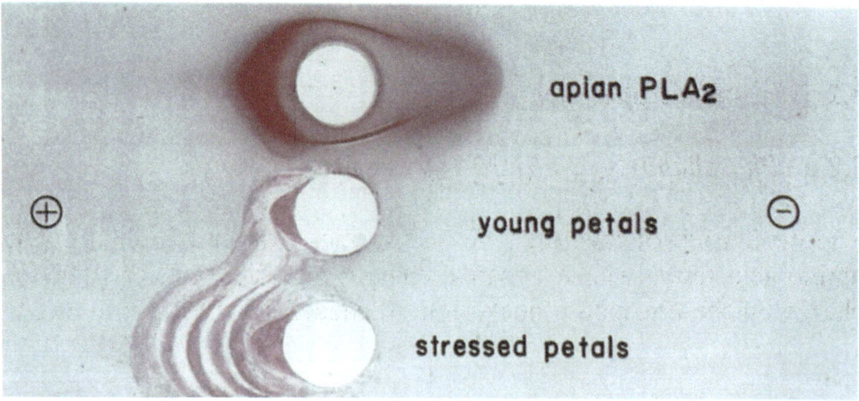

Colorplate III. Identification of plant phospholipase A_2 in rose petals by "Rocket" immuno-electrophoresis (see pg. 177).

tered (McRae and Thompson, 1983). This observation suggests that changes in protein conformation may be an essential facet of the aging process. The proteins that apparently are involved are discussed herewith.

Basic research conducted at the Max Planck Institute in Germany by the group of Deisenhoffer and Michel led to the unravelling of the complex structure of the photosynthetic reaction center in the photosynthetic bacterium *Rhodopseudomonus viridis*. This center (cf. Colorplate I.b.) is comprised of four protein subunits – L, M, H and cytochrome c – which are covalently bound to a *heme* group. In addition to these basic building blocks, each reaction center contains four chlorophyll molecules, two phaeophytin molecules, two quinone groups and one divalent Fe^{2+} molecule which is not a part of the heme nucleus. The reader interested in further details of the stereochemistry of the photosynthetic reaction center of the above photosynthetic bacterium (which in higher plants is assumed to be similar, if not identical) is referred to computer-aided color diagrams based on crystallographic data published by Deisenhoffer et al. in 1985 in Vol. 318 of *Nature* – a publication, whose scientific import led to the granting of the 1988 Nobel Prize to this group.

5.13. References and recommended reading

Aragoncillo, C., Sanchez-Monge, R. and Salcedo, G. 1981. Two groups of low molecular weight hydrophobic proteins from barley endosperm. *Jour. Exp. Bot.* 32:1279–86.

Arondel, V., Tchang, F., Vergnolle, C., Jolliot, A. Grosbois, M., Guerbette, F., Morch, M.D., Pemollet, J.C., Delsany, M., Puidoménech, P. and Kader, J.C. 1989. Plant lipid transfer proteins: structure, function and molecular biology. pp. 341–50. In: Eds. P.A. Biacs, K. Gruiz and T. Kremmer. *Biologlcal Role of Plant Lipids*. Akademiai Kiado, Budapest and Plenum Press, New York.

Barbier-Brygoo, H., Ephritikhine, G., Klämbt, D. Gluslain, M. and Gucrn, J. 1989. Functional evidence for an auxin receptor at the plasmalemma of tobacco mesophyll cells. *Proc. Nat. Acad. Sci. USA* 86:891–5.

Bretscher, M.S. 1985. The molecules of the cell membranes. *Sci. Am.* 253:86–90.

Brinegar, A.C., Cooper, G., Stevens, O., Hauer, C.R., Shabanowitz, J., Hunt, D.F. and Fox, J.E. 1988. Characterization of a benzyladenine binding site peptide from a wheat cytokinin-binding protein. *Proc. Nat. Acad. Sci. USA* 85:5927–31.

Chadwick, C.M. and Garrod, D.R. (Eds.) 1986. *Hormones, Receptors and Cellular Interactions in Plants*. Cambridge Univ. Press.

Coleman, H.A. 1986. Cl^- currents in *Chara* – a patch clamp study. *J. Membr. Physiol.* 93:55–61.

Cosgrove, D.J. and Hedrich, R. 1990. Stretch activated cation, anion and calcium channels ω-exist in guard cells. *Plant Physiol.* 93:suppl.93.

Deckman, M., Haimovitz, R. and Shinitzky, M. 1985. Selective release of integral proteins from human erythroyte membranes by hydrostatic pressure. *Biochim. Biophys. Acta* 821:334–40.

Deisenhoffer, J., Epp, O., Muki, K., Huber, R. and Michel, H. 1985. Structure of the protein sub-units in the photosynthetic reaction centre of *Rhodopseudomonas viridis* at 3 Å resolution. *Nature* 318:618–624.

Emons, A.M.C. and Traas, J.A. 1986. Coated pits and coated vesicles on the plasma membrane of plant cells. *Eur. J. Cell Biol.* 41:57–64.

Finkelstein, A. 1987. *Water Movement Through Lipid Bilayers, Pores and Plasma Membranes. Theory and Reality*. John Wiley and Sons, New York.

Folke, L.C., Edwards, K.L., Pickard, B.G. and Misler, S. 1988. A stretch activated anion channel in tobacco protoplasts. *FEBS Lett.* 237:141–4.

Gahan, P.B. 1984. *Plant Histochemistry and Cytochemistry: An Introduction*. Academic Press, London.

Gahan, P.B. 1989. Viability of protoplases. pp. 36–9. In: Ed. Y. Bajaj. *Biotechnology in Agriculture and Forestry*. Springer Verlag, Berlin–Heidelberg.

Grosbois, M., Guerbette, F., Donady, D. and Kader, J.C. 1987. Enzyme immunoassay of a plant phospholipid transfer protein. *Biochim. Biophys. Acta* 917:162–8.

Guern, J. 1987. Regulation from within: the hormone dilemma. *Ann. Bot.* 60:75–102.

Harper, J.F., Surowy, T.K. and Sussman, M.R. 1989. Molecular cloning and sequence of cDNA encoding the plasma membrane proton pump (H^+-ATPase) of *Arabidopsis thaliana*. *Proc. Natl. Acad. Sci. USA* 86:11234–8.

Hassidim, M., Braun, Y., Lerner, H.R. and Reinhold, L. 1990. Na^+/H^+ and K^+/H^+ antiport in root membrane vesicles isolated from the helophyte *Atriplex* and the glycophyte, cotton. *Pl. Plys.* 94:1745–1801.

Hedrich, R. and Schroeder, J.I. 1989. The physiology of ion channels and electrogenic pumps in higher plants. *An. Rev. Plant Physiol. Plant Mol. Bio.* 40:539–69.

Hedrich, R., Stoeckel, H. and Takeda, K. 1990. Electrophysiology of the plasma membrane of higher plant cells: new insights from patch clamp studies. pp. 182–202. In: Eds. C. Larsson and I.M. Møller. *The Plant Plasma Membrane*. Springer Verlag, Berlin–Heidelberg.

Horn, M.A., Heinstein, P.F. and Low, P.S. 1989. Receptor mediated endocytosis in plant cells. *Plant Cell.* 1:1003–9.

Hornberg, C. and Weiler, E.W. 1984. High affinity binding sites for abscisic acid on the plasmalemma of *Vicia faba* guard cells. *Nature* 310:321–4.

Huang, H., Yang, S.Y. and Tang, Y.W. 1990. A functional cytokinin binding protein in photochemical reaction of chloroplasts. pp. 173–8. In: Eds. M. Pharis and S. Rood. *Plant Growth Substances 1988*. Elsevier, Amsterdam.

Jayabaskaran, C. 1990. Isolation and characterization of a cytokinin binding protein from cucumber cotyledons. *Plant. Gr. Reg.* 9:9–18.

Jones, A.M. 1990. Do we have an auxin receptor yet? *Physiol. Plant.* 80:154–8.

Jones, C.P. and Paleg, L.G. 1984. Complex formation between IAA and phospholipid components in aqueous media. 2. Interaction of auxins and related compounds with phosphatidyl membranes. *Biochemistry* 23:1521–4.

Kader, J.C. and Laurent, P. 1990. Lipid synthesis and transfer. pp. 314–23. In: Eds. Nijkamp, L.H.W. van der Plas and J. van Aartrijk. *Progress in Plant Cellular and*

Molecular Biology. Kluwer Academic Publishers, Dordrecht, The Netherlands.

Kader, J.C., Ostergaard, J., Vergnolle, C. and Renard, M. 1991. Bifunctional lipid transfer/fatty acid binding proteins in plants. pp. 272–4. In: Eds. P.J. Quinn and J.L. Harwood. *Plant Lipid Biochemistry, Structure and Function*. Portland Press, London and Colchester.

Katekar, G.F. 1979. Auxins: on the nature of the receptor site and molecular requirements for auxin activity. *Phytochemistry* 18:223–33.

Kikugawa, K. and Beppu, M. 1987. Involvement of lipid oxidation products in the formation of fluorescent cross-linked proteins. *Chem. Phy. Lipids* 44:277–96.

Klämbt, D. 1987. *Plant Hormone Receptors. NATO ASI series*. Springer Verlag, Berlin.

Landau, E.M. and Leshem, Y.Y. 1988. Biophysical interactions of membrane anionic phospholipids with pH, calcium and auxins. *J. Exp. Bot.* 39:551–9.

Larsson, C., Møller, I.M. and Widell, S. 1990. An introduction to the plant plasma membrane – its molecular composition and organization. pp. 1–15. In: Eds. C. Larsson and I.M. Møller. *The Plant Plasma Membrane*. Springer Verlag, Berlin–Heidelberg.

Lear, J.D., Wasserman, Z.R. and deGrado, W.F. 1988. Synthetic amphiphilic peptides for protein ion channels. *Science* 240:1177–81.

Leshem, Y.Y. 1987. Ca^{2+} and intermolecular bridging of membranal phospholipids and proteins. pp. 225–7. In: Eds. P. Stumpf, J. Mudd and W. Ness. *The Metabolism, Structure and Function of Plant Lipids*. Plenum Press, New York.

Leshem, Y.Y. and Bar-Nes, G. 1987. Hormone receptor manipulation by hydrostatic pressure: interaction between Ca^{2+} membrane components and phosphatidylinositol in pea foliage microsomes. pp. In: Ed. D. Klämbt. *Plant Hormone Receptors. NATO ASI series*. Springer Verlag, Berlin.

Leshem, Y.Y., Cojocaru, M., Margel, S., El-Ani, D. and Landau, E.M. 1990. A biophysical study of abscisic acid interaction with membrane phospholipid components. *New Phytol.* 116:487–98.

Ling, G.N. 1988. A physical theory of the living state: application to water and solute distribution. *Scanning Microscopy* 2:899–913.

Ling, G.N. 1991. *A Revolution in Physiology of the Living Cell*. R.P. Kriger Publications (In press).

Löbler, M. and Klämbt, D. 1985. Auxin binding protein from coleoptile membranes of corn. Localization of a putative receptor. *Jour. Biol. Chem.* 260:9854–9.

Low, R.L., Terlin, B.S. Schauf, C.L. and Stockton, M.E. 1990. Ca^{2+} activation of *Mougeotia* potassium channels. *Plant. Physiol.* 92:822–30.

Maasthuis, F.J.M. and Prins, H.B.A. 1990. Patch clamp studies on root cell vacuoles of a salt bolerant and a salt sensitive *Plantago* species. *Plant Physiol.* 92:23–8.

Manwaring, J.D. and Csallany, A.S. 1988. Malondialdehyde-containing proteins and their relationship to Vitamin E. *Lipids* 23:651–4.

McCarty, D.R., Carson, C.B., Stinard, P.S. and Robertson, D.S. 1989. Molecular analysis of *viviparous*-1: an ABA-insensitive mutant of maize. *Plant Cell* 1:523–32.

McRae, D. and Thompson, J.E. 1983. Conformational changes in the thylakoid proteins correlate with a decline in photosynthetic electron transport during senescence. *Proc. Can. Soc. Plant Phys. Abstr.* 73:28.

Mellor, J.R. and Sanson, M.S.P. 1990. Ion channel properties of mastoparan – a 14 residue from wasp venom and of MI3 – a 12 residue analogue. *Proc. R. Soc. London* B239:383–400.

Mennes, A.M., Boot, C.J.M., Mann, A.C., Libbenga, K.R. and van der Zaal, E.J. 1990. IAA perception and auxin regulated gene expression. pp. 100–5. In: Eds. P. Pharis and S.B. Rood. *Plant Growth Substances 1988*. Springer Verlag, Berlin.

Mennes, A.M., Nakamura, C., van der Linde, P.C.G., van der Zaal, E.J., van Zelgen, H.J., Quint, A. and Libbenga, K.R. 1987. Cytosolic and membrane-bound high-affinity auxin binding proteins in tobacco. *Jour. Biol. Chem.* 260:51–62.

Meyer, T., Holowka, D. and Stryer, L. 1988. Highly cooperative opening of calcium channels of IP_3. *Science* 240:653–6.

Moran, M., Ehrenstein, G., Iwasa, K., Bare, C. and Mischke, C. 1984. Ionic channels in plasmalemma of wheat protoplasts. *Science* 226:835–8.

Moran, W. 1990. Stretch cultivated channels in plasmalemma of pulvinar motor cells. *Plant Physiol.* 93:supl.91.

Moreau, P., Bertho, P., Juguelin, H. and Lessire, R. 1988. Intracellular transport of very long chain fatty acids in etiolated leek seedlings. *Plant. Physiol. Biochem.* 26:173–8.

Morré, D.J., Brightman, A.O., Wu, L.Y., Barr, R., Leak, B. and Crane, F.L. 1988. Role of plasma membrane redox activities on elongation growth in plants. *Physiol. Plant.* 73:187–93.

Morris, C.E. 1990. Mechano-sensitive ion channels. *Jour. Mem. Biol.* 43:93–107.

Napier, R.M. and Venis, M.A. 1990. Receptors for plant growth regulators: recent advances. *Jour. Plant Growth. Reg.* 73:113–26.

Oleski, N.A., Mahdavi, P. and Bennet, A.B. 1987. Transport properties of the tomato fruit tonoplast. II. Citrate transport. *Plant Physiol.* 74:997–1000.

Pantoja, O., Dainty, J. and Blumwald, E. 1990. Tonoplast ion channels from sugar beet cell suspensions. *Pl. Phys.* 94:1788–94.

Paz-Ares, J., Ponz, F., Aragoncillo, C., Hernandez-Lucas, C., Salcedo, G., Carbonero, P. and Garcia-Olmedo, F. 1983. In vivo and in vitro synthesis of CM-proteins (A-hordeins) from barley (*Hordeum vulgare L.*). *Planta* 157:74–80.

Pennel, R.I., Knox, J.P., Scofield, G.N., Selvendran, R.R. and Roberts, K. 1989. A family of abundant plasma-membrane associated glycoproteins related to the arabinogalactan proteins is unique to flowering plants. *J. Cell Biol.* 108:1967–77.

Pick, U. and Racker, E. 1979. Purification and reconstitution of the N,N′-dicyclohexyl-carbodiimde-sensitive ATPase complex from spinach chloroplasts. *J. Biol. Chem.* 254:2793–9.

Popot, J.L. 1987. La structure des proteins membranaires. *La Recherche* 192:1170–82.

Reiss, H.D. and Herth, W. 1985. Nifedipine-sensitive calcium channels are involved in polar growth of lily pollen tubes. *Jour. Cell. Sci.* 70:247–54.

Reithmeir, R.A.F. 1985. Assembly of proteins into membranes. pp. 503–58. In: Eds. D.E. Vance and J.E. Vance. *Biochemistry of Lipids and Membranes*. Benjamin Cummings, California.

Rickers, J., Spener, F. and Kader, J.C. 1985. A phospholipid transfer protein that binds long chain fatty acids. *FEBS Lett.* 180:29–32.

Robinson, D.G. and Hillmer, S. 1990. Coated pits. pp. 233–55. In: Eds. C. Larsson and I.M. Møller. *The Plant Plasma Membrane*. Springer Verlag, Berlin–Heidelberg.

Rodriguez-Loperena, M.A., Aragoncillo, C., Carbonero, P. and Garcia-Olmedo, F.

1975. Heterogeneity of wheat endosperm proteolipids (CM proteins). *Phytochem.* 14:1219–23.

Romanov, G.A., Taran, V.Y., Chvojka, L. and Kulaeva, O.N. 1988. Receptor like cytokinin binding protein from barley leaves. *J. Pl. Gr. Reg.* 7:1–17.

Salcedo, G., Rodriguez-Loperena, M.A. and Aragoncillo, C. 1978. Relationship among low MW hydrophobic proteins from wheat endosperm. *Phytochem.* 17:1491–4.

Satter, R.L. and Moran, N. 1988. Ionic channels in plant cell membranes. *Physiol. Plant* 72:816–20.

Saunders, J.J. 1986. Cytokinin activation and redistribution of plasma-membrane ion channels in *Funaria*. *Planta* 167:402–9.

Schatz, G. 1986. *The Biogenesis of Membranes*. Biocenter. University of Basel. pp. 12–13.

Schauf, C.L. and Wilson, K.J. 1987. Properties of single K^+ and Cl^- channels in *Asclepias tuberosa* photoplasts. *Plant Physiol.* 85:413–41.

Schroeder, F. 1985. Use of fluorescence spectroscopy in assessment of biological membrane properties. pp. 193–217. In: Eds. R. Aloia, C. Curtain and L. Gordon. *Advances in Membrane Fluidity. I.* Alan-Liss Inc., New York.

Schroeder, J.I., Raschke, K. and Neher, E. 1987. Voltage dependence of K^+ channels in guard cell protoplasts. *Proc. Nat. Acad. Sci. USA* 84:4108–12.

Schwitzguebel, J.P. and Siegenthaler, J.P. 1985. Evidence for the lack of phospholipid transfer protein in the stroma of spinach chloroplasts. *Plant Science* 40:167–71.

Shewry, P.R., Lafiandra, D., Salcedo, G., ARagoncillo, C., Garcia-Olmedo, F., Lew, E.J.L. and Kasarda, D.D. 1984. N-terminal amino acid sequences of chloroform/methanol-soluble proteins and albumins from endosperms of wheat, barley and related species. *FEBS Lett.* 175:359–63.

Shih, C.Y., Dumbroff, E.B. and Thompson, J.E. 1989. Identification of a naturally occurring inhibitor of the conversion of ACC to ethylene by carnation microsomes. *Plant Physiol.* 89: 1053–9.

Shimomura, S., Sotobayashi, T., Futai, M. and Fukui, T. 1986. Purification and properties of an auxin binding protein from maize shoot membranes. *Jour. Biochem.* 99:1513–24.

Singer, S.J. and Yaffe, M.P. 1990. Embedded or not? hydrophobic sequences and membranes. *TIBS* 15:369–73.

Sisler, E.C. 1990. Ethylene binding sites in plants. Is there more than one? pp. 192–200. In: Eds. M. Pharis and S.B. Rood. *Plant Growth Substance 1988.* Springer Verlag, Berlin.

Sisler, E.C. and Goren, R. 1981. Ethylene binding. The basis for hormone action in plants. *What's New in Plant Physiol.* 12:37–40.

Smith, P.G., Venis, M.A. and Hall, M.A. 1985. Oxidation of ethylene by cotyledon extracts from *Vicia faba*. Cofactor requirements and kinetics. *Planta* 163:97–104.

Srivastava, L.M. 1987. The gibberellin receptor. pp. 199–227. In: Ed. D. Klämbt. *Plant Hormone Receptors. NATO ASI Series. 10.* Springer Verlag, Berlin–Heidelberg.

Steer, M.W. 1988. Plasma membrane turnover in plant cells. *Jour. Exp. Bot.* 39:987–96.

Stoeckel, H. and Takeda, K. 1990. *Calcium activated, voltage dependent, non selective cation currents in endosperm plasma membrane from higher plants.* Proc. Royal Soc. London B. (In press).

Sussman, M.R. and Harper, J.F. 1989. Molecular biology of the plasma membrane of higher plants. *Plant Cell* 1:953–60.

Tchang, F., This, P., Stiefel, V., Arondel, V., Morch, M.D., Pagès, M., Puigdoménech, P., Grellet, F., Delseny, M., Bouillon, P., Huet, J.C., Guerbette, F., Beauvais-Canté, F., Duranton, H., Pemollet, J.C. and Kader, J.C. 1988. Phospholipid transfer protein: full length cDNA and amino acid sequence in maize. *J Biol. Chem.* 263:16849–55.

Thompson, J.E. 1983. Personal communication.

Venis, M.A. 1985. *Hormone Binding Sites in Plants*. Longman, Harlow.

Venis, M.A., Napier, R. and Bollon, M.A. 1990. Membrane bound auxin receptors. pp. 106–12. In: Eds. R. Pharis and S. Rood. *Plant Growth Substances 1988.* Springer Verlag, Berlin.

Waneck, G., Stein, M. and Flavell, A. 1988. Conversion of a PI-anchored protein by a single amino acid mutation. *Science* 241:697–9.

Widell, S. and Larsson, C. 1990. A critical evaluation of markers used in plasma membrane research. pp. 17–43. In: Eds. C. Larsson, I.M. Møller. *The Plant Plasma Membrane*. Springer Verlag, Berlin–Heidelberg.

Widger, W.R., Cramer, W.A., Herrmann, R.G. and Trebst, A. 1984. Sequence homology and structural similarity between cytochrome of mitochondrial complex mitochondrial III and the chloroplast b6-f complex: position of the cytochrome hemes in the membranes. *Proc. Natl Acad. Sci. USA* 81:674–8.

Yamada, M., Tsuboi, S., Osafune, T., Suga, T. and Takishima, K. 1991. Multifunctional properties of non-specific lipid transfer protein from higher plants. pp. 278–80. In: Eds. P.J. Quinn and J.L. Harwood. *Plant Lipid Biochemistry. Structure and Utilization*. Portland Press, London and Colchester.

6. Membrane water

Besides lipids, proteins, sterols and carbohydrates, membranes also contain another component that is often ignored – water. The association with, and distribution of, water in the membranes is not uniform and aqueous domains differing in location and physico-chemical properties may be found on the membrane's exterior and at different sites in the membrane's interior.

6.1. Extra-membrane water

It has been shown that the average membrane bilayer thickness *is approximately 20–30 Å greater than the distance taken between the headgroups on each side*. This leads to the conclusion that a membrane is bilayer sandwiched between two water layers each 10–15 Å thick each of which contributes to permeation phenomena (Disalero and de Gier, 1983). The total amount of hydration depends on lipid constituency, phase and temperature, as well as upon hydrophilicity of membrane proteins. Phosphatidylcholine in the liquid-crytalline (L-α) phase has 15 moles of water per mole of lipid, while in the gel state it has less than 8 (Cevc, 1981). Hydration moveover depends upon type of phospholipid headgroups – phosphatidylcholine ones being more hydrated than those of phosphatidylethanolamine. At this point it is worthy of mention that water can be replaced at bilayer interfaces by sugars such as sucrose and trehalose (Crowe and Crowe, 1986).

Concerning the contribution of the typical plant membrane-contained sugar – *galactose* – to membrane hydration, there is an interesting situation, depending upon galactose location. The overall trend as exemplified in chloroplastic MGDG and DGDG is for galactose to increase hydrophilicity most markedly and it may well be that the high degree of hydration endowed by these species contributes to the initial water availability for cleavage in the process of photosynthesis. However when galactose is aligned in the particular tightly packed configurations encountered in sterolglycosides or glycosylceramides (detailed in Chapters 3 and 4) the hydration capacity is markedly reduced to only 1 mole of H_2O per mole of lipid.

The degree of membrane hydration dependent upon the above described factors has relevance to overall plant sensitivity to temperature and water stress. The following table indicates comparative hydrophilicities of membrane components:

Table 6.1. Degree of hydration of various plant membrane lipid constituents

Membrane lipid component	Moles H_2O/mole lipid
Glucosylceramide	1
Sterolglycosides	1
PC (L β-gel phase)	8
PEA	<15
PC (L α-liquid crystalline phase)	15
MGDG	17
DGDG	34

6.2. Vicinal or solvated water

A second type of membrane-contained water is termed *vicinal* (after the Latin *vicinalis* meaning neighboring) or *solvated water*. This term is applied to water located within the bilayer and which is attracted to, or in approximation with the expanses of membrane constituents, be they fatty acyl sidechains, proteins, sterols or carbohydrates. These water molecules so situated are associated with atoms of the constituents of the surface itself with the result that they become restricted in their motion. Because of its particular structure, vicinal water has distinct biophysical properties: it is about 3% less dense than bulk water, has a greater heat capacity, and its viscosity is higher than bulk water. Solvation forces also affect several layers that are not immediately adjacent to the contacting interfaces. If the surface itself contributes to increasing structuring of neighboring water molecules, these molecules similarly affect the next layer and so on, for some distance. Estimations varying between 10–16 molecular layers from the surface have been suggested. The amount of vicinal or solvated water, which may overlap with certain amounts of the other two aforementioned membrane water categories, has been estimated to include 90% of the total (Drost-Hansen and Singleton, 1989). It seems that the hydration pattern of all common glycerophospholipids is quite similar. The polar groups of the lipid heads that are most strongly effected are the phosphate, carboxyl, carbonyl, and choline residues. Bilayer hydropholicity, therefore, depends strongly on the participation of these groups in interlipid hydrogen bonds since these influence the availability of the headgroup proton donors and acceptors for hydrogen bonding with the water molecule (Cevc and Marsh, 1987). As to the interior of the membrane in certain interstices between the individual monolayer leaflets of the bilayer "*trapped*" *water* which, prevents its crystallizing even at −20°C, is also to be found (Lafleur et al., 1989; Meier et al., 1990).

6.3. Bulk water

The misnomer, at least in terms of membrane structure, of *bulk water* is used to describe water that exists in an essentially unchanged physical state and while apparently only comprising a relatively low proportion of membrane water is a major transporter of ions, assimilates, and other solutes. Basing on both theoretical and experimental findings, the molecular distance between individual glycerolipid membrane constituents (ca 4 Å) has been calculated to be greater than the radius of an individual H_2O molecule (ca 2 Å). This provides the major (but not sole) conduit of trans-membrane water transport via a straightforward osmotic/diffusion process.

In bulk water in its natural state the constituent water dipole molecules are by no means completely randomized – at 20°C as much as 85% of the molecules are still in a quasicrystalline state as compared to 100% hydrogen binding in ice. The nature of hydrogen binding is such that if it occurs between molecules A and B, it continues between B and C, and from C proceeds to D and so forth until a closed group of H_2O molecules, termed a *cluster*, is formed (see Fig. 6.1). These clusters form and disperse subject to localized fluctuations in energy levels and the picturesque term employed to describe this dynamic process is – "*flickering clusters*". This situation may be compared to a folk-dance in which many couples participate and periodically, following a cue in the melody, change partners. At any given moment, to continue the simile, 85% of the individuals have partners, while later the same percentage but with different partners may be observed. The lifetime of a flickering cluster in bulk water is ca 10^{-11} seconds,

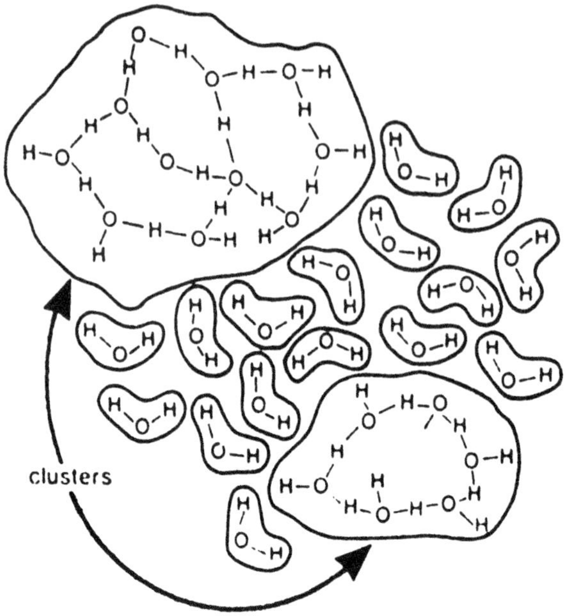

Fig. 6.1. "Flickering clusters" in bulk water.

and cluster size *decreases* with *increase* of temperature. Kosmotropes (see next section) increase size and stability of clusters while chaotropes destabilize the structure. As shown by magnetic resonance imaging, relaxation of bound to free state of water contributes to winter dormancy break in deciduous fruit tree buds (Faust et al., 1991).

6.4. Solute effects on the physical state of water

In terms of effects on the physical state of water which has direct bearing on responses of the membrane to physiological changes, solutes may be divided into two classes each having an opposite effect (Hatefi and Henstein, 1969; Oku and McDonald, 1983; Koynova et al., 1989):

a. *Chaotropic solutes*
b. *Kosmotropic solutes*

Chaotropic (derived from the Greek *chaos*) *reagents* decrease interactions between apolar molecules by destabilizing the quasi-crystalline structure of bulk water and in the membrane favor micellar phases and less the laminar ones. Apolar groups form hydrophilic bonds mainly as a result of their thermodynamically unfavorable interaction with water, rather than as a consequence of interaction with each other. The transfer of an apolar molecule from the lipophilic surrounding to water is endergonic by 2–6 kcal per mole because of an associated unitary entropy decrease of 10–20 entropy units. This large negative entropy appears to be almost entirely related to the structure of water and can be diminished by changing the structure of water in the direction of greater disorder. In other words chaotropes give rise to regions in which water structure is broken down, melted or depolarized as compared to pure water. There is also a tendency to make water *more* lipophilic and moreover a greater propensity to induce lipid oxidation in membranes. A series of chaotropes in decreasing order of effectivity has been found as follows:

$$\text{tribromoacetate} > \text{trichloroacetate} > ClO_4 > SCN^- > NO_3^-$$

Kosmotropic (from the Greek *kosmos*, order) *solutes* or reagents on the other hand stabilize the quasicrystalline state of bulk water, which even at 20°C is 85% in an ordered state; thus doing these kosmotropes tend to decrease the area of contact between membrane phospholipids and the intervening bulk water. Koynova et al. (1989) have found that membrane-associated sugars favor the formation of hexagonal (H_{II}) phase in membranes at the expense of laminar liquid crystalline (L-α) phases and in this respect act as kosmotropic reagents. Based upon the above, these workers reject the "water replacement" interpretation of cryoprotection (Crowe and Crowe, 1986) which assumes that sugars can substitute for water bound to membrane surfaces. At the same time no claim is made that sugars do not act as endoprotectants or as protectants under untoward conditions of dessication – the mode of action involved is believed to be a kosmotropic one. Leshem et al. (1990) have suggested that the senescence-promoting and stomate-closing effect of the plant hormone *abscisic acid* may be attributed to a kosmotropic effect induced in the water adjacent to a membrane phospholipid or to a *lyso*-phospholipid.

Anion effects which dominate the classical Hofmeister effect in terms of chaotropic and kosmotropic properties have been categorized as follows (Collins and Washabaugh, 1985):

$$SO_4^{2-} \simeq HPO_4^{2-} > F^- > Cl^- > BI^- > I^- \simeq ClO_4^- > SCN^-$$

Those anions which are before Cl^- are polar kosmotropes and they stabilize proteins, while those listed after Cl^- are chaotropes and by breaking water structure destabilize proteins. Cl^- itself has little effect on water structure nor, in the range of 0.1–9.7 M, on protein stability. The above Hofmeister series as mentioned previously is dominated by anion effects and becomes physiologically important at moderate concentrations – in the 0.01–1.0 M range.

The above applies to anions. Concerning cations the following sequence is followed:

$$Mg > Ca > Li > Na \mid K > Rb > Cs$$

The ions on the left represent, in decreasing order, the kosmotropic bulk water stabilizing effect, while those on the right act in the opposite manner.

6.5. Water channels (?) and movement across membranes

This subject in all its complexity has been described in a brief volume (Finkelstein, 1987) and the approach presently taken follows the leitmotifs there sounded but sometimes with a varying interpretation.

The pathway of water across cell membranes has been a central topic both in plant and mammalian physiology. It is also a topic that from its inception has often combined, often in a none too harmonious manner, biological and physical disciplines. The early research conducted on osmosis by the plant physiologist Pfeffer in 1877 ten years later stimulated the physical chemist van't Hoff to interpret osmotic equilibrium in a physical theory. Since then and until the present day physiologists, biophysicists and physical chemists have provided equally pertinent theories for non-equilibrium situations. The pioneering efforts of Kedem and Katchalsky (1962) of the Weizmann Institute, Rehovoth, Israel provided a firm basis for a thermodynamic interpretation of osmosis, thus opening a vista of industrial application of "*reverse osmosis*" which is now employed for large scale water desalination and in medical dialysis biotechnology for combatting kidney malfunction. Disalero (1988) has extended Kedem and Katchalsky's concepts to include further details for permeation of water through lipid membranes and has stressed interfacial and *elastic* phenomena.

Finkelstein has aptly pointed out that the essentially multidisciplinary approach has remained a mixed blessing since physiologists' interests in the theory of water transport have not been motivated by a desire to fathom the intricacies of physico-chemical theory. Their interest has rather been of how application of water permeability measurements can afford a means of addressing important questions regarding the pathway of water and solute movement through cell membranes. A pivotal question that may be posed is – does water movement in and out of cells occur through the bilayer proper of the plasma membrane, or through aqueous pores in the membrane, or through both?

Water has a finite solubility in hydrocarbon, and the water permeability of lipid bilayer membranes is adequately explained by the hydrocarbon-like properties of the bilayer interior. Moreover, since the values of the water permeability coefficients, which are dependent on the lipid composition of the bilayer, span a range that includes most of those reported for cell membranes, there is no difficulty in accounting for the large magnitudes of the water permeability coefficients of plasma membranes simply in terms of the properties of their lipid bilayer matrices. However the existence of ion-conducting channels and other transport systems in plasma membranes provides additional potential pathways for water movement that cannot be dismissed a priori. Indeed, it is almost certain that water is permeant through most of these regions, and therefore the question is a quantitative one: which is the *predominant* route that water utilizes in crossing plasma membranes? Is it through the lipid bilayer, or through polar transport sites (pores)? Related to this question is the problem of establishing criteria for experimentally distinguishing between these two possibilities. A further question is, if indeed a pore mechanism is located in the membrane, then may ion-conducting channels fulfill this function or at least contribute to the water movement pathway?

Answers to these questions have been partially provided and the overall picture that is emerging is that water transport occurs primarily by a solubility diffusion process through the lipid bilayer of the plasma membrane. However in a few cell membranes it has clearly been concluded that water permeation is via *pores* – this notably being the case in the red blood cell. Quantitative calculations have shown that water transport ionic channels in the very few investigated cases cannot account for the major part of water transport. Moreover it was found that the water pores of the red blood-cell membrane are virtually ion impermeable. To the best of the author's knowledge the possibility of water channel existence has as yet not been addressed in plant membranes.

Certain antibiotics such as nystatin, amphotericin B and gramidicin A are known to induce formation of water channels in membranes with which they come into contact, this water-channeling capacity in some cases may, and in others may not, be connected with their bacteriocidal properties. Nystatin and amphotericin B differ from gramadicin A in that for water channel formation they have an absolute requirement for presence of sterols in the membrane (Graziani and Livne, 1972) thus limiting their efficiency as therapeutic agents against bacteria whose membranes lack sterols.

Meanwhile, despite Finkelstein's elegant contentions, general opinion holds that the existence of specific water channels in biological membranes remains a moot issue, and in the case of plant membranes still is *Terra incognita* inviting exploration.

6.6. References and recommended reading

Cevc, G. 1981. Water and membranes:the interdependence of their physico-chemical properties in the case of phospholipid headgroups. *Stud. Biophys.* 91:41–6.

Cevc, G. and Marsh, D. 1987. *Phospholipid Bilayers: Physical Principles and Models.* Wiley and Sons, New York.

Collins, K.D. and Washabaugh, M.W. 1985. The Hofmeister effect and the behaviour of water at interfaces. *Quart. Rev. Biophys.* 18:323–422.

Crowe, J.H., Crowe, L.M. and Chapman, D. 1984. Preservation of membranes in anhydrobiotic organisms – role of trehalose. *Science* 223:701–63.

Crowe, L.M. and Crowe, J.H. 1986. Hydration dependent phase transitions and permeability properties of biological membranes. pp. 210–30. In: Ed. A.C. Leopold. *Membranes, Metabolism and Dry Organisms.* Cornell Univ. Press, Ithaca.

Disalero, E. 1988. Permeability of water and polar solutes in lipid bilayers. *Adv. Coll. Interf. Sc.* 29: 141–70.

Disalero, E.A. and de Gier, J. 1983. Contribution of aqueous interphases to the permeability barrier of lipid bilayers for non-electrolytes. *Chem. Phys. Lipids* 32:39–47.

Drost-Hansen, W. and Singleton, L. J. 1989. Liquid asset – how exotic properties of cell water enhance life. *The Sciences*, Sept–Oct edition: 38–42.

Faust, M., Liu, D., Millard, M.M. and Stutte G.W. 1991. Bound versus free water in dormant apple buds – a theory for endodormancy. *HortSci.* 26:887–90.

Finkelstein, A. 1987. *Water Movement Through Lipid Bilayers, Pores and Plasma Membranes. Theory and Reality.* John Wiley and Sons, New York.

Graziani, Y. and Livne, A. 1972. Water permeability of bilayer lipid membranes: sterol-lipid interaction. *J. Membrane Biol.* 7:275–84.

Hatefi, Y. and Henstein, 1969. Solubilization of particulate proteins and non-electrolytes by chaotropic reagents. *Proc. Nat. Acad. Sci. USA* 1129–36.

Kedem, O. and Katchalsky, A. 1962. A physical interpretation of the phenomenological coefficients of membrane permeability. *J. Gen. Physiol.* 45:143–79.

Koynova, R.D., Tenchov, B.G. and Quinn, P.J. 1989. Sugars favour formation of hexagonal phase at the expense of lamellar liquid crystalline phase in hydrated phosphatidylethanolamines. *Biochim. Biophys. Acta* 980:377–80.

Lafleur, M., Pigeon, M., Pezolet, M. and Caillé, J.P. 1989. Raman spectrum of interstitial water in biological systems. *Phys. Chem.* 93:1522–6.

Leshem, Y.Y., Cojocaru, M., Margel, S., El-Ani, D. and Landau, E.M. 1990. A biophysical study of abscisic acid interaction with membrane phospholipid components. *New Phytol.* 116:487–98.

Meier, E.M., Schummer, D. and Sandhoff, K. 1990. Evidence for the presence of water within the hydrophobic core of membranes. *Chem. Phys. Lip.* 55:103–13.

Oku, N. and McDonald, R.C. 1983. Solubilization of phospholipids by chaotropic ion solutions. *J. Biol. Chem.* 258:8733–8.

Part two

Biophysical aspects of membrane structure and function

7. Plant membrane biophysics in development and senescence

Based essentially upon a biophysical approach but couched in terminology understandable to the average plant biologist the present chapter will highlight some of the current issues of membrane biophysics which at present are commanding considerable attention in the field of both developmental and senescence physiology.

7.1. Membrane phases

Current opinion holds that the bilayer is not the sole phase present in any given membrane, nor are the various membrane categories uniform in their phase composition and that within a certain membrane species different phases may co-exist. Following is a brief discussion of the major phase types.

Membrane glycerolipids (phospho-, galacto- and sulpholipids) appear in the following phases (see also Fig. 7.1):

A. Bilayer configurations
1. Liquid-crystalline or fluid phase (L-α)
2. Solid or gel phase (L-β)

B. Non-bilayer configurations
1. Hexagonal (H_1)
2. Inverted hexagonal (H_{II})
3. Cubic (I_I)
4. Inverted cubic (I_{II})

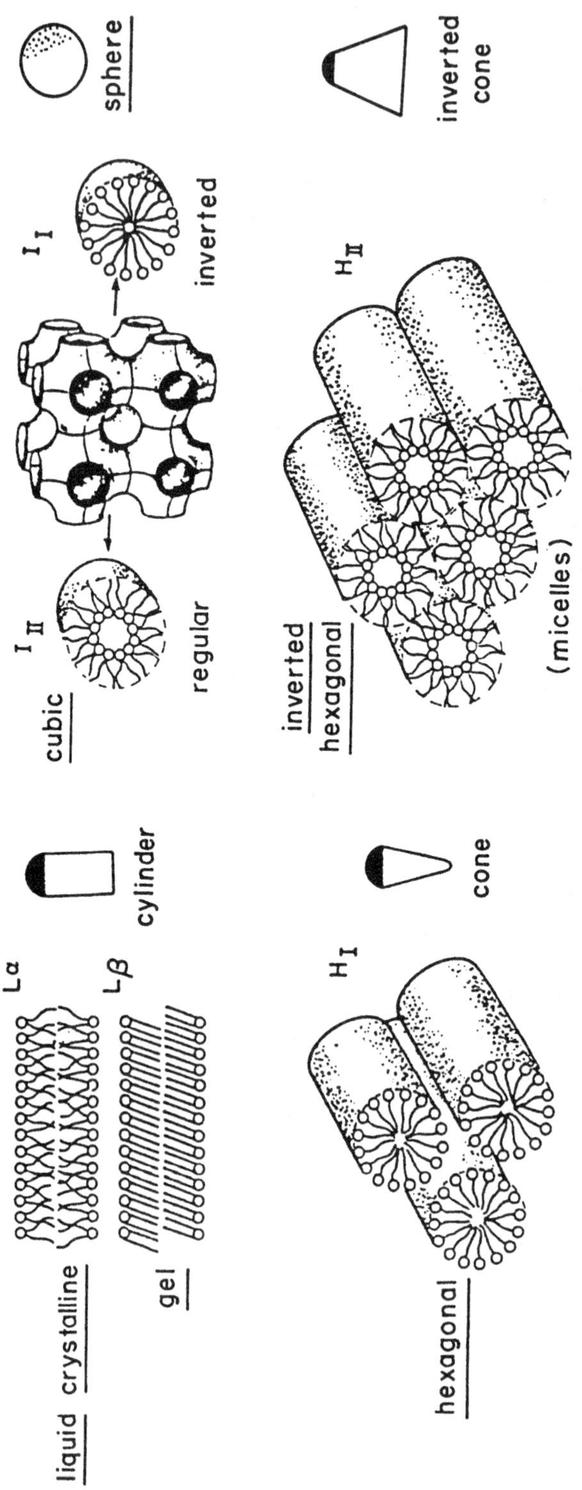

Fig. 7.1. Major membrane phase configuration. The figures appearing to the right of each phase scheme represent the geometric shape of the individual PL units.

C. Mixed configurations
In such membranes, for example, certain domains may assume a L-α bilayer configuration in which other phases may be interspersed.

7.1.1. *The liquid crystalline or fluid phase*

This phase is typical of young functioning cells. The membrane bilayer is comparatively pliable, since the fatty acid chains possess motional freedom, and as a result of this the embedded proteins, like buoys in a lipid ocean, manifest maximal biological activity. The high motional freedom of the phospholipids has been measured in synthetic membranes where each molecule is able to exchange positions with its neighbor(s) 10^6 times per second. Preponderance of unsaturated fatty acids (linoleic, linolenic and arachidonic acids) enhances the fluid phase, since as stated before, the *cis*-double bonds in these fatty acid chains induce 'kinks' which increase pliability. In this context it has been claimed that one of the mechanisms of thermophilic adaptation is a relative increase in content of such bonds.

7.1.2. *The gel or solid phase*

In this phase, typical of senescing or faulty membranes, the phospholipid fatty tails in the bilayer are 'frozen' and completely lose their motional freedom. As a result the membrane becomes more rigid and forfeits its pliability, and the embedded proteins are no longer able to move. In the event of physical stress this frozen state precludes membrane adaptation; as a result, leaks are formed and overall membrane permeability and function are impaired. Gel phase membranes are also more prone to phospholytic enzyme action. The phase transition from the liquid crystalline to the gel is therefore considered a foreboding omen and a forerunner of senescence and finally, of death.

Typical adverse manifestations of membrane phase transitions have been reported during loss of pollen viability (Crowe et al., 1989), avocado fruit chilling (Platt-Aloia and Thompson, 1987) and pea foliage senescence (Leshem et al., 1984; Sridhara and Leshem, 1986). The temperature at which this phase change occurs, designated *transition temperature* (Tc) will be discussed presently.

7.1.3. *The hexagonal phases*

Current models of membranes postulate a bilayer in which the lipid component serves as a matrix within which proteins fulfill specialized functions. However, as shown by De Kruijff et al. (1980) and de Gier et al. (1982), the model in which lipids are regarded as inert building blocks is not sufficient to account for all membrane phenomena. Under certain circumstances, including temperature transitions, the increase in intercellular divalent ion concentrations which could occur during senescence, and a certain degree of tissue hydration, amongst other factors, may induce non-bilayer *hexagonal* I or II configurations or their inclusion as sections of the bilayer (see Figs 7.1, 7.2 and Plate 7.1). The hexagonal II phase, in which the phospholipid molecules form long cylinders with the polar headgroups facing a water-filled core, is one of the most common of these non-bilayer configurations and possibly includes single-tailed *lyso*-phospholipids. Interactions between adjacent phospholipid cylinders are hydrophobic. The diameter of the water-filled core and the center-to-center distance of the phospholipid cylinders depend on the hydration of the tissue, the presence of various cations and the type of phospholipid. These non-bilayer structures may play dynamic roles in diverse membrane-mediated processes, including fusion and the ability to withstand dehydration, as occurs for example in seeds. It has been suggested but as yet not experimentally proved that such inverted micelles may serve as endogenous calcium ionophores or channels.

Concerning the H_I hexagonal phase, it is believed that it is formed as a result of the production of *lyso*-phospholipids i.e. PLs with only one PL tail (see Fig. 7.1). The regulatory effect of *lyso*-PLs has recently been demonstrated in plants (Palmgren and Sommarin, 1989). Moreover Leshem (1991) has shown in terms of surface chemistry that *lyso*-PL as compared to PL has a marked decreasing effect of surface tension of PC monolayers; this possibly simulating the loss of the saturated acyl *sn*-2 chain by phospholipase A_2-like action during membrane senescence (see also p. 20 in Chapter 2).

7.1.4. *The cubic phases*

The cubic phases somewhat resemble the hexagonal ones as seen in Fig. 7.1 but differ in that spheres, and not cylindrical micelles are formed. The term cubic could be misleading but should be understood in the light of the regular geometrical block-like pattern obtained

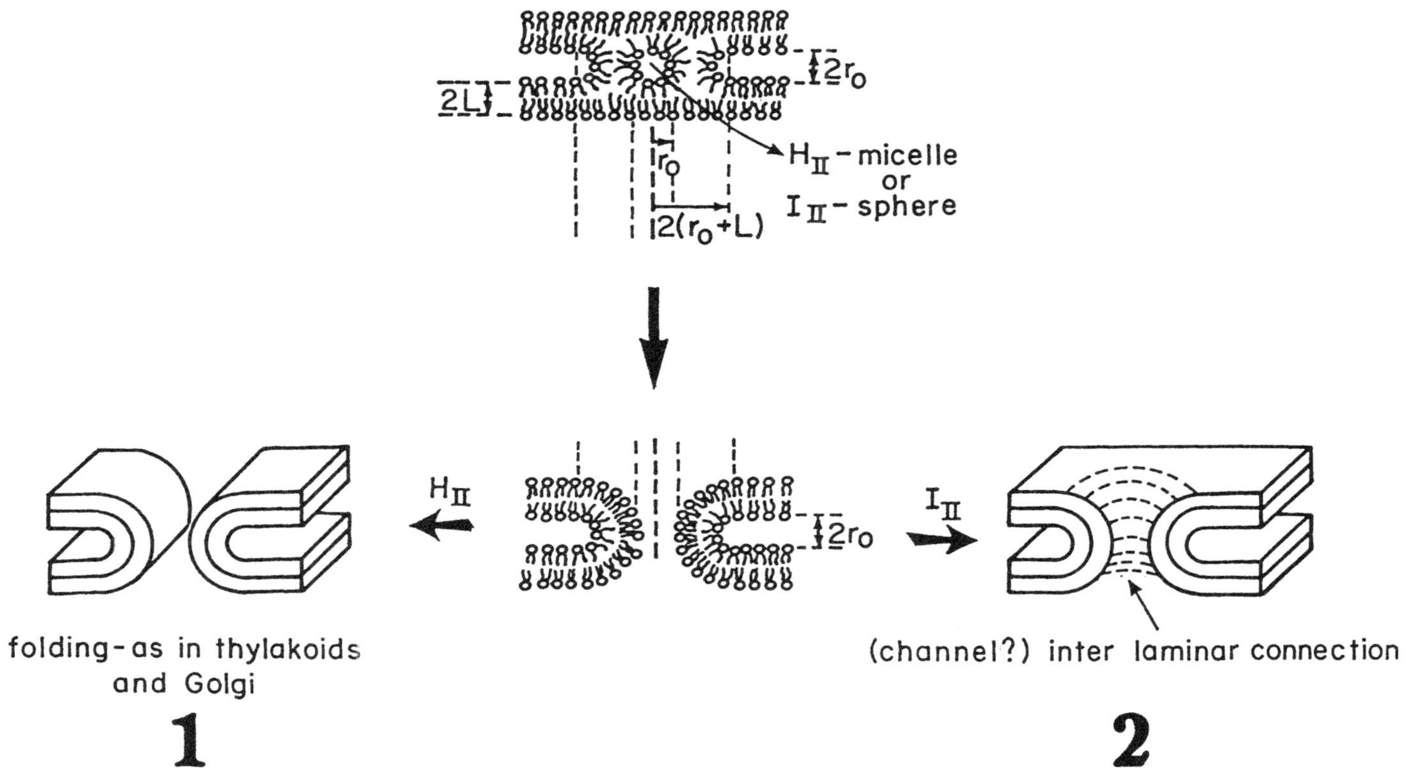

Fig. 7.2. Diagrammatic representation how curvature and connections may be obtained between two proximal membrane bilayers. Modified from Siegel (1986).

Plate 7.1. The inverted hexagonal (H_{II}) phase in bean thylakoid membranes. The area in the center of the electron micrograph indicates the inverted micelles forming the H_{II} phase over part of the membrane. Bar line represents 0.1 μm. Plate reproduced with the courtesy of Prof. P.L. Quinn of the University of London.

upon stacking of the units (see Fig. 7.1). Recently it has been demonstrated that the inverted cubic phase I_{II} apparently is an intermediary formation during the $L\alpha$ and H_{II} phase transition as could occur during stress or ageing:

$$L\alpha \rightarrow I_{II} \rightarrow H_{II}$$

According to Skyamsander et al. (1988) the I_{II} configuration is stable between the temperature range of 2°C–15°C for a period of several weeks. As demonstrated in Figs 7.1 and 7.2 the I_{II} configuration may also participate in the formation of highly curved regions of biological membranes as occurring in thylakoids and Golgi apparatus as well as in forming of intra membrane pores, coated pits and channels (Siegel, 1986; Leshem et al., 1990): see Fig. 7.2. In this context Lindblom and Rilfors (1989) have reported that the preponderance of the galactolipid MGDG, which can comprise ca 50% of chloroplastic membranes, favors I_{II} mediated formation of the typically curved chloroplastic membranes. Rawyler and Siegenthaler (1989) have reported that the photosynthesis inhibiting herbicide DCMU blocks electron transport

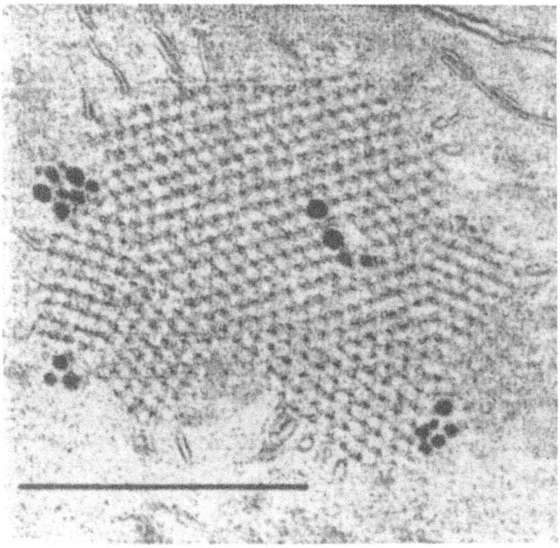

Plate 7.2. Electron micrograph of a prolammelar body (PLB) from an etioplast of wheat. The structure of this PLB is of the zincblende type. The bar represents 1 μm. Reproduced with the courtesy of Dr. Eva Selstam, University of Umea, Sweden.

by detrimentally altering compaction and curvature of folded thylakoid membranes: this implies some interference with cubic phase formation.

7.1.4.1. *The cubic phase and the prolammelar body*
In leaves of dark grown plants proplastids differentiate to etioplasts. In this plastid two different membrane systems are formed, the normal planar prothylakoid membrane and the regularly branched crystalline prolamellar body (PLB) membrane (Plate 7.2). As mentioned in Chapter 1 the biological membrane has a barrier function that would be destroyed if the membrane lipids were arranged in a non-bilayer structure. However, the PLB in the etioplast of dark grown plants is proposed to be a cubic phase structure formed by thylakoid lipids and one major protein. The formation of a cubic phase by the etioplast inner membrane probably is not a disadvantage, since the PLB membrane has no barrier function, but rather functions as a storage place for the chlorophyll precursor protochlorophyllide and the thylakoid lipids. Upon being exposed to light the protochlorophyllide is reduced to chlorophyllide and the cubic PLB membrane is rearranged to a bilayer. The transformation from the cubic bicontinuous structure to the normal bilayer does not require any large rearrangements, since

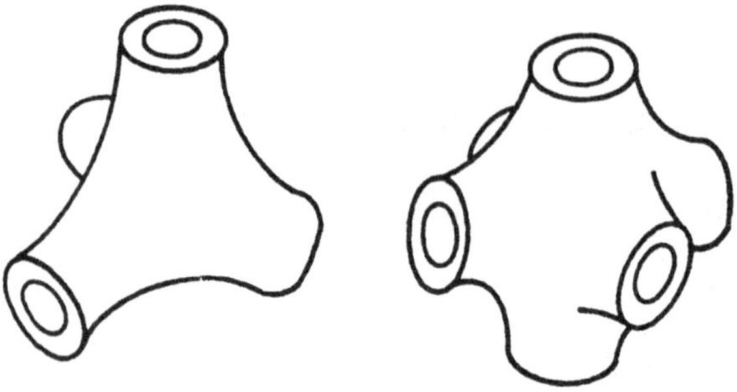

Fig. 7.3. 4- and 6-armed units of the tetrapodal and cubic PLB structures (After Selstam et al., 1990).

the lipids in the cubic phase already are in a bilayer configuration (Lindstedt and Liljenberg, 1990; Selstam et al., 1990, 1991).

Meticulous electron microscopy and other considerations have revealed that the PLB membrane is a unit membrane where the crystalline lattice is formed by 4- or 6-armed units as depicted in Fig. 7.3 (Selstam et al., 1990). The 4-armed unit is a tetrapod where 4 tubes meet in one point with equal angles. Tetrapods are arranged in different networks analogous to the crystal structures of wurtzite and of zincblende (= diamond). The PLB can be formed by either the wurtzite or of the zincblende type, or by a combination of the two types in one PLB. The 6-armed unit is a cubic structure where the 6 tubes meet at right angles.

The crystalline structure in both the tetrapodal and the cubic PLB are similar to the structures formed by infinite periodic minimal surface (IPMS). IPMS is defined to have a mean curvature of zero. Curvatures described by minimal surfaces have the free energy of the curvature at a minimum. Several IPMS structures have been described – the D-type (diamond) is formed by tetrapodal units and is similar to the PLB structure of zincblende type (Lindstedt and Liljenberg, 1990) while the P-type (primitive) has a cubic symmetry which is similar to the 6-armed PLB structure.

7.1.5. *Mixed phases*

The occurrence, within a given membrane, of different domains each having a separate phase configuration typical of senescing membranes has far-reaching physiological implications. The interface of such domains may induce "leaks" in membrane structure, this portending cell death. These interfaces may also be sites of attachment of the senescence-evoking plant hormone abscisic acid (Stillwell et al., 1987). Moreover if one individual leaflet undergoes a phase change, embedded intrinsic proteins may experience steric disorientation (Schroeder, 1984). On the other hand the ordered and arranged co-existence of different membrane phases may in certain cases be advantageous (Quinn, 1989): a typical case being the thylakoid membrane. Here occurrence of molecular glycerolipid species which do not assume bilayer configurations (MGDG and PE) together with those that do, is of prime importance for photosynthesis.

Methods of determining phase state of membranes vary. In plants wide angle X-ray diffraction and freeze fracture electron microscopy have often been used. These methods are discussed in a later section of this chapter. Changes in phase are related to membrane microviscosity and this is the topic of the next section.

Transition temperatures (*T*c): One of the commonly employed means to determine phase state of a biology is by assessing its *transition temperature* which may be defined as the lowest temperature at which initial signs of a phase change from the liquid-crystalline to the solid-gel can be detected. A high *T*c is regarded as physiologically more stress sensitive than a low one. As a rule *T*c increases with age, environmental stress and chilling sensitivity. Moreover at a given age plants differ in tropical and sub-tropical species possessing a high *T*c whereas deciduous or cold resistant ones have a low one (Pauls and Thompson, 1982; Leshem et al., 1986; Vigh et al., 1987). A further example is the behavior of microsomal membranes derived from bean foliage where normal *T*c was 18°C, but increased to 30°C after ozone treatment (Pauls and Thompson, 1980).

It has been found that in most species of subtropical fruit *T*c's are within the range of 10°C–15°C whereas in deciduous fruit which do not suffer from chilling, the *T*c range is 3°C–5°C. The break in an Arrhenius plot which on one axis presents the logarithm of some physiological parameter (e.g. respiration) and on the other the reciprocal of the absolute temperature in Kelvin degrees, is often considered to correspond to the *T*c of the tested tissue or fruit. This is a comparatively easy-to-perform assay and hence its widespread use in "scientific" horticulture. Fig. 7.4 presents a comparison of *T*c of tomatoes which are not cold-resistant to that of cauliflower inflorescences which even at temperatures below zero are unharmed.

It should however be mentioned that this "easy" method to determine *T*c is not always considered reliable and that it only serves as a general indication, as is the relationship of a temperature below which it is not desirable to store fruit to *T*c as established by this method.

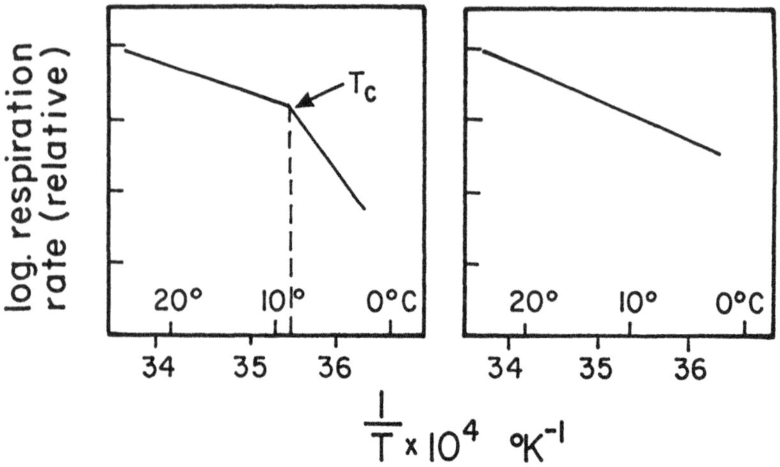

Fig. 7.4. Arrhenius plots of transition temperatures (*T*c) of tomatoes (left) and cauliflower inflorescences (right). It can be seen that at 6°C tomato membranes undergo a phase change from $L_\alpha \rightarrow L_\beta$ which corresponds to the break in this plot. Cauliflower inflorescences even below zero manifest no break.

This inter alia stems from the fact that while phase transition does indeed occur at the "break-point" of the Arrhenius plot, cold-damage to the fruit or vegetable is manifested much later with a delay of weeks, days or even months.

The phase transition principle in post-harvest fruit physiology has been described for avocado fruit by Platt-Aloia and Thompson (1987). It was found that exposing the fruit to a combination of a lower temperature (6°C) and ethylene caused a phase change from the liquid-crystalline to the gel: moreover until a certain stage this phase change was reversible viz. when the avocado fruits were returned to higher temperatures, phase change from gel to liquid-crystalline was observed. In contrast, after a prolonged exposure to a temperature below 6°C, detrimental phase transition occurred and the *T*c of the tested avocado fruits was irreversible with resulting irreparable damage to the stored produce.

Following are some of the factors which may affect phase transition of glycerolipid bilayers:

a) the length of fatty acyl sidechains;
b) their degree of unsaturation and location. In plant systems, the lipid phase of leaf membranes shifts from fluid to gel phase (phase separation) during senescence, but changes in fatty acid saturation in membranes of aged leaves are of insufficient magnitude to account for an accompanying rise in transition temperature. However a selective loss of phospholipid results in an augmented neutral-

lipid to phospholipid ratio. The lipid extracts of chloroplast membranes reveal that neutral lipids formed during senescence order the phospholipid into the gel phase;

c) the type of headgroup, and especially headgroup's anionic charge;
d) the extent of Ca^{2+}-mediated electrostatic binding;
e) the presence of phytohormones, especially auxins and cytokinins.

Concerning "b" above – location and saturation of fatty acyl tailgroups – it has interestingly been pointed out that when an unsaturated fatty acyl tail is at the *sn*-2 site in PC, it has a greater *Tc* decreasing effect than when located at the *sn*-1 site (Davis et al., 1981; Coolbear et al., 1983). At this stage it should be remembered that in many a plant membrane there is a very heterogeneous array of fatty acid species in PC and other membrane phospholipids, this leading to a rather wide *Tc* range.

Wide angle X-ray diffraction and membrane phase transitions: Assessment of phase situation can be accomplished by several techniques amongst which *wide angle X-ray* diffraction is one of the most commonly employed. Although wide angle X-ray diffraction analysis is a highly sophisticated technique for studying the physical structure of membranes, its drawback is that it does not provide specific information regarding the chemical changes which occur in relation to alteration of physical structure. Wide angle X-ray diffraction patterns reveal the Bragg spacings which only reflect the distance between the fatty acid side chains. This distance between fatty acid side chains is influenced by chemical substrates. But it is important to realize that the fluidity could be influenced by either lipid-lipid interactions, lipid-protein interactions or protein-protein interactions as well as by electrostatic charge.

Plate 7.3 depicts X-ray diffraction patterns of senescing pea foliage microsomal membranes of young (right) and senescent (left) plant membranes. In the L-α liquid crystalline phase typical of intact and young membranes the motional freedom of the lipid chains produces a diffuse diffraction ring pattern, the center of which in Bragg spacing is 4.6 Å from the periphery. In the L-β gel phase the fixed, non-motional chains produce a clearly demarcated X-ray diffusion band whose Bragg spacing is 4.2 Å from the periphery. Hexagonal and bilamellar phases with hexagonal sections are assessed by NMR or by freeze-fracture electron microscopy.

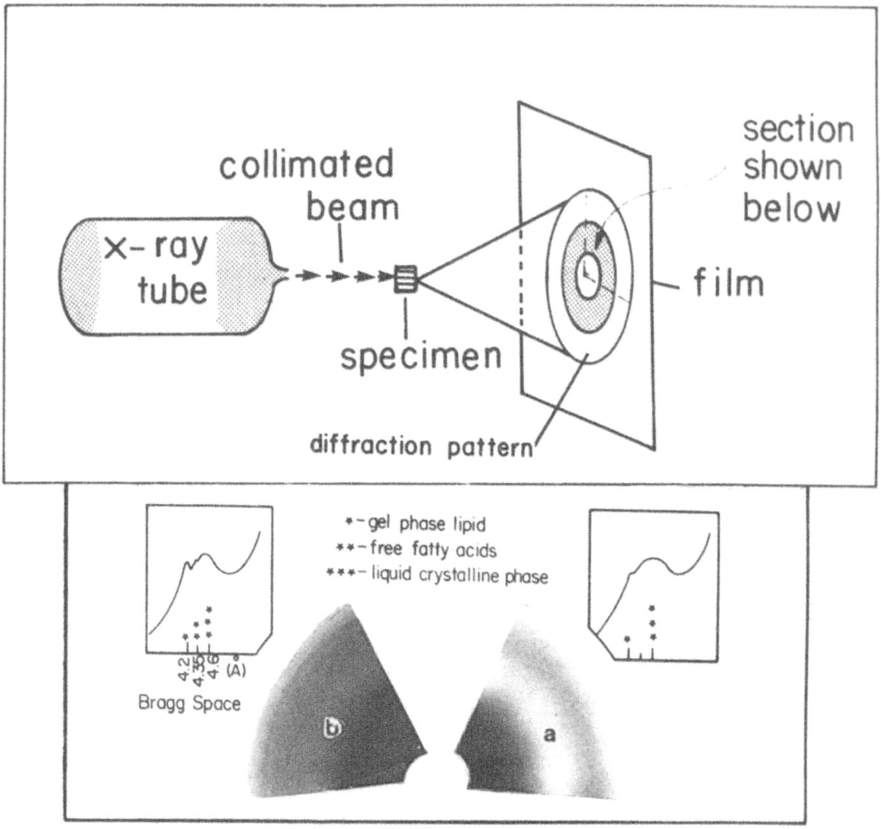

Plate 7.3. Wide angle X-ray diffraction patterns of microsomal membranes in bean cotyledons. *above*: X-ray diffraction method. *below*: *a. Young membranes*; Note the diffuse halo on the periphery for young membranes. This indicates a fluid, crystalline phase of the lipid bilayer with low probability of 'leak' formation. Bragg spacing of this diffuse band centers around 4.6 Å. *b. Senescent membranes*; In the senescent membranes, note the appearance of a sharply demarcated halo beyond the diffuse band. This 'hard edged' halo indicates that regions of the membrane have assumed the gel state. In this phase membranes lose flexibility, many become leaky, and embedded proteins are unable to move. Bragg spacing of this gel band centers around 4.2 Å. The wide diffuse 10 Å band in the center is a reflection which has not been well characterized but is thought to be derived from a protein (After Leshem et al., 1986).

7.2. Membrane fluidity

7.2.1. *Microviscosity* (η)

As pointed out by Shinitzky and Barenholz (1978) who originally coined the term *microviscosity* (expressed in *poise* units), ageing in a wide variety of mammalian species and organs within a given species is invariably accompanied by *increased* η, or conversely by *decreased* fluidity of membranes.

The dynamics of cell surface membranes are mainly determined by the fluid state of the lipid layer while rotational and translational mobility and degree of exposure of membrane proteins are mostly determined by the fluidity of the lipid layer (Shinitzky and Inbar, 1974). Thermal motion of membrane components designated *membrane fluidity* is considered to play an important role in cellular recognition mechanisms. The mobility of receptor sites on the surface membrane may provide a mechanism for conduction of specific biochemical signals to the cell's interior, whereas the fluid state of the surface membrane lipid layer may control signals that relate grossly to the flexibility of the membrane and the thermal mobility of receptors embedded in it. The reader wishing to pursue the topic of biological membrane fluidity in general is referred to the comprehensive volumes edited by Shinitzky (1984) and by Aloia (1985).

Plant studies on η in various plant organs including fruit, flowers, foliage and seed cotyledons have clearly indicated that loss of membrane fluidity (the inverse of η) is associated with both normal and imposed ageing. The increase of viscosity is believed to cause a marked curtailment of both *lateral* and *rotational diffusion* (detailed later in this chapter) of embedded proteins. Resultingly their biological activity is decreased because of a lower statistical probability of an enzyme to encounter its substrate, or a receptor its agonist. Moreover the viscous state reduces membrane pliability and *leaks* may be formed. Increase of rigidity of cerebral cortex membranes has also been found to be associated with human ageing. Rigidification of plasma membranes has an overall effect of curtailing signal transduction.

Membrane fluidity depends upon a wide array of both endogenous and exogenous factors – the former include fatty acyl desaturation, Ca^{2+}-cross linking, pH; while the latter encompass changes in micro- or macroenvironmental stress involving temperature and salinity, etc. One school of thought, further detailed in the chapter dealing with membranes and cold responses, claims that environmental conditions especially pertaining heat and cold, involve a process of *homeoviscous adaptation*. In this process plants strive to sustain an optimal membrane fluidity (either under thermophilic conditions or when subjected to low temperature regimes) by adaptively changing the degree of saturation or desaturation of glycerolipid fatty acyl chains. According to this concept, which is not without its challengers, under high temperatures, degree of saturation *increases*, the converse applying to low temperatures. The active agents in such homeoviscous adaptation is claimed to be the *desaturase* family of enzymes (Sinenski, 1974; Raison et al., 1982). This process of homeoviscous adaptation or conversely, the lack of ability of membranes to adapt in such a manner has been claimed to be one of the factors determining whether a plant species may be an annual or a perennial (Pike, 1982).

Be this as it may in plant membrane acclimation to vicissitudes of temperature, the mechanism of varying fatty acid unsaturation or hydrogenation (hardening) is a trick of margarine food technology where in order to prevent melting of produce in summer, triacylglycerol saturation is somewhat increased, while in winter to enable easy spreading over bread etc. unsaturation is increased.

An often used experimental tool in plant membrane fluidity studies is *melittin*, a polypeptide component of bee venom which is a potent inhibitor of photochemical reactions in chloroplasts. Melittin intercalates with chloroplast membrane lipids and reduces fluidity.

7.2.2. *Assessment of microviscosity (η) by fluorescent depolarization*

Fluorescent in vivo probing of biological membranes, numerical assessment of which is entered into the empiric Shinitzky-Barenholtz (1978) formula enables calculation of η which is expressed in *poise* units.

The empirical Shinitzky-Barenholzformula is as follows:

$$\eta = \frac{2P}{0.46 - P} \text{poise} \qquad [1]$$

In this equation, η = microviscosity; P is the degree of fluorescence depolarization.

The most commonly used probe is *diphenyl hexatriene* (DPHX) – see Fig. 7.5. Refinement of the DPHX probing technique (Lentz, 1989) differentiates between the following DPHX species:

- *DPHX* per se – aligns parallel to the PL fatty acyl chains and is a measure of η of the internal section of the bilayer.

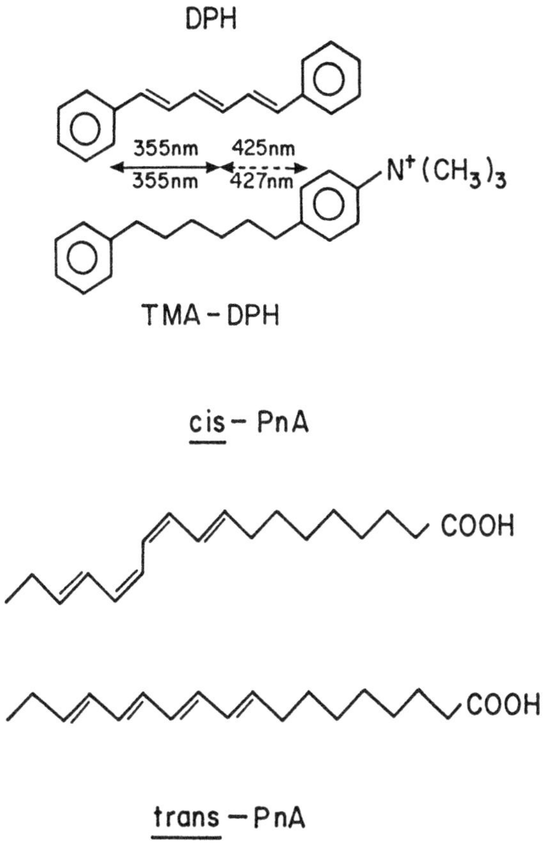

Fig. 7.5. Molecular structure of commonly used fluorescent membrane probes *above* – diphenylhexatriene (DPHX) and trimethylamine DPHX (—— excitation, ------ emission wavelength). *below* – parinaric acid (PnA) which is a lipid analog anionic probe.

- *DPHX-p-phosphatidylcholine* – the PC moiety of this probe associates with the outer hydrophilic domain of the bilayer while the DPHX associates with the inner hydrophobic domain: thus a better overall assessment of η is obtained.

  While this DPHX-based technique was initially developed for mammalian membranes, it has been effectively applied to various plant membranes (McKersie et al., 1976; Leshem and Inbar, 1978; Pauls and Thompson, 1982).

  Data obtained from a wide variety of biological organisms clearly indicate that senescence and stress are accompanied by a marked *increase* of membranal microviscosity. A typical example is given in Fig. 7.6 which shows that with progress of senescence of rose petal plasmalemmas there is a progressive increase in membrane microviscosity.

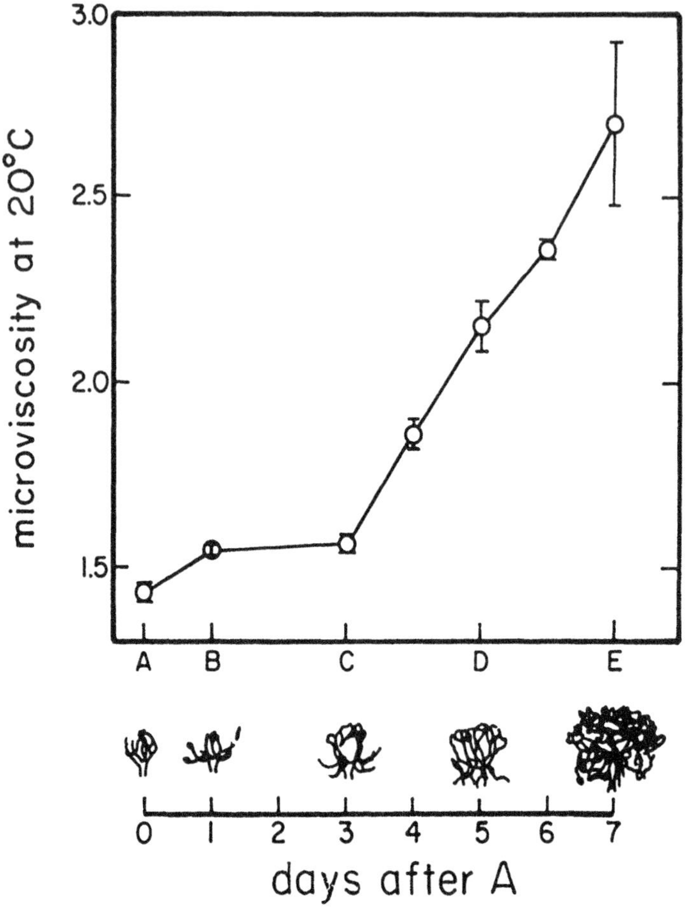

Fig. 7.6. The increase of rose-petal plasmalemmal viscosity with age (Borochov et al., 1978).

- *TMA* (*trimethylamine*) – DPHX. This probe assumedly is positioned like DPHX-p-PC and also presents an overall and not a localized η pattern. Climacteric of tomato pericarp plasma membrane tissue has been reported to coincide with increase of steady state anisotropy as detected by this probe (Legge et al., 1985). In *Vicia faba* foliage plasma membranes TMA-DPHX proved to be a far more sensitive means of monitoring abscisic acid mediated increase of rigidity than DPHX (Leshem et al., 1990). Plate 7.4 and Colorplate II indicate the topographical specificity of this fluorescent probe.

The drawback of DPHX derivatives, as compared to DPHX are their limited penetrability: however if an appropriate solvent solution is employed (Lentz, 1989) this can be overcome.

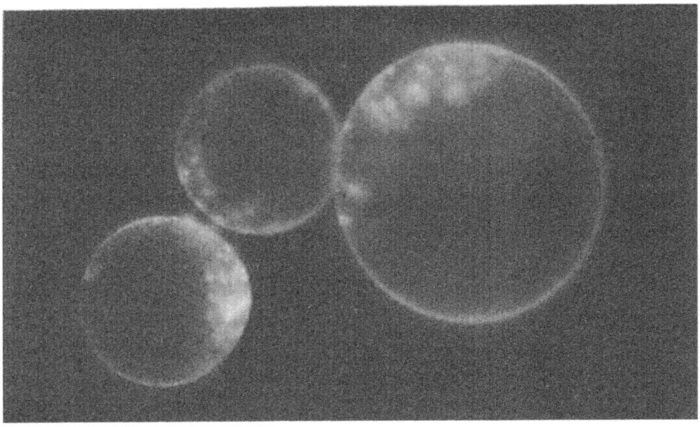

Plate *7.4.* TMA-diphenylhexatriene fluorescent probing of the plasma-membrane of rye protoplasts. The bright fluorescence on the periphery indicates probe location. Photograph kindly supplied by Dr. Robert Langis, Department of Botany, McGill University, Quebec, Canada. (See also colorplate II page 96).

Other fluorescent probes include:

- Dansyl-phosphatidyl ethanolamine (DPE) which is a membrane phospholipid headgroup probe.
- *cis* and *trans parinaric acid* (PnA) – These are among the few non-aromatic fluorophores that are lipid analogs (see Fig. 7.5). *Cis*-PnA partitions between liquid crystalline and gel phase lipid while *trans*-PnA enters the lipid domain (Sklar, 1980; Legge et al., 1985).

Fluorescence probing pertains to membrane lipids whose diffusional movements are rapid. However when assessing the particular contribution of membrane proteins, the principle of *phosphorescence* is employed since phosphorescent probes when excited have a far longer half-life than fluorescence probes and therefore are more suited to the tardier diffusional protein movements. In this case it is necessary to employ a light "chopper" which differentiates between phosphorescence and fluorescence.

7.2.3. *Factors effecting* η

1. Length and degree of unsaturation of fatty acyl chains (Brenner, 1984; Leshem et al., 1988; Leshem, 1989).

   In this respect a most interesting trend is followed (before reading the following sentences it is recommended that the reader turn back to Fig. 2.3) – with the increasing content of the listed fatty acyl chains changes are as follows:

   $18:0 \rightarrow 18:1$ – marked change in η

18:1 → 18:2 – change effected, but less than above
18:2 → 18:3 – negligible change.

This can be understood in that the degree of fatty acyl chain bending increases from 18:0 → 18:2 after which it begins to turn back upon itself.

2. Nature of the glycerolipid headgroups especially the intensity of the negative charge of the anionic species $PIP_3 > PIP_2 > PI > SQDG$. PC and PEA have a net neutral charge (Landau and Leshem, 1988; Larsson et al., 1990 – see also Table 3.3).
3. Degree of divalent ion especially Ca^{2+} mediated electrostatic bridging of adjacent phospholipid molecules (Leshem, 1987). In this context cell membrane integrity of stressed apple fruit tissue has been reported to be dependent on divalent cations (Nur et al., 1986).
4. Effect of phytohormones. These may act directly on PL components (Rojas and Tobias, 1965; Jones and Paleg, 1984; Leshem et al., 1990), or on interfaces between different phase domains (discussed under the following section), or indirectly by means of hormone complexing with membrane-contained receptors. The general pattern which is emerging is that auxin and cytokinin appear to fluidize membranes, gibberellin, like sterols, stabilizes them and abscisic acid rigidifies them.

In conclusion to this section, it is stressed that while η increase of membranes is a common factor accompanying plant ageing, not all membrane types are equally sensitive; the mitochondrial double bilayered membrane exhibits a certain degree of enhanced stability and this may account for the observation that in senescing plant cells, the mitochondrion is the most resistant organelle to ripening-associated cellular disruption (Leshem and Inbar, 1978).

7.3. Surface tension parameters and membrane function

From a biological point of view a membrane may be regarded as a system in which, because of their relative lightness, two lipid monolayers float upon an aqueous medium which is the cytosol. In surface chemistry the latter is designated the *subphase*. The assembly may be likened to a bowl of chicken broth in which the main constituent, the aqueous section, is covered by a layer of fatty drops – with the small difference in that in the cell the fatty layer is usually a layered continuum and not present as droplets. In this context the two bio-

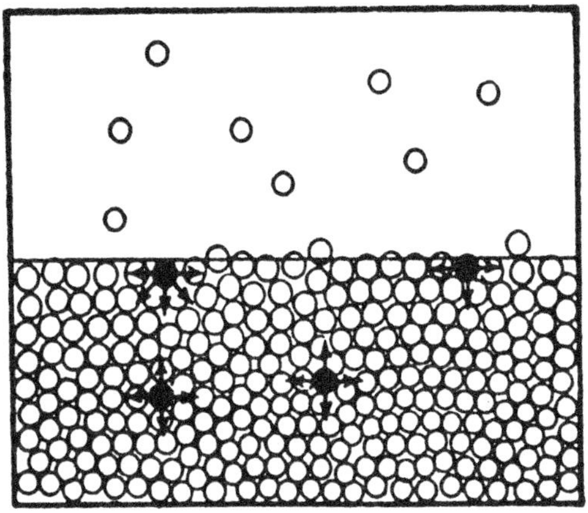

Fig. 7.7. The formation of a physically different film over the surface of a liquid at a liquid:air interface. It is noted that the *anistropy* of the molecules located on the surface results from force applied, as here graphically depicted, only from three directions, while a like liquid molecule situated deeper within the liquid is subjected to equal forces from all sides this resulting in an *isotropic* situation.

physical properties that exert far-reaching effects on the membrane are *surface tension* and *surface pressure*.

7.3.1. *Surface tension* (γ)

This parameter is familiar to the reader who has observed how a needle floats or an insect walks on water. The film formed over the water:air interface is the result of unequal or *anisotropic* cohesive forces acting on the water molecules located on the liquid's surface. As illustrated in Fig. 7.7 this anisotropic situation at the surface forms a more tightly packed layer of water molecules. The physical forces here differing from the *isotropic* forces acting equally on all sides of a given water molecule deeper within the aqueous medium.

Water possesses an exceptionally high surface tension, 72.35 mN m^{-1} at 24°C, which is relatively stable and varies by <3% between 5°C and 30°C. However, addition of amphiphiles such as phospholipids (PL) at the air/water interface considerably reduces the surface tension and since surface tension phenomena at aqueous cytosol/PL interfaces in membranes may contribute to biological regulation, the present section endeavors to outline how certain biological factors may affect this parameter. Mariners of old when pouring oil on stormy waters decreased surface tension and resultingly, the height of sea waves. The spherical shape of a water drop is also due to surface tension.

7.3.2. *Surface pressure* (π)

This parameter is the converse of the former and is the result of the thermo-dynamic action of liquid particles at the liquid:air interface and also of the anisotropic situation of the liquid molecules at the surface layer. As seen in Fig. 7.7 a given H_2O molecule on the surface is subjected to equal "bombardment" by its neighbors on either side, this *not* being the case on a perpendicular plane where the pressure exerted from below is not counteracted by pressure from above. Resultingly the physical momentum acting upwards causes an overall net monolayer pressure designated *surface pressure* which is usually expressed in the same units as surface tension – milliNewtons/meter ($mN\,m^{-1}$) or in the older terminology – dynes/meter.

Under experimental conditions it is possible to calculate surface pressure (π) from surface tension (γ) in monolayers by employing the following formula:

$$\pi = \gamma(H_2O) - \gamma(H_2O + \text{monolayer}) \qquad [2]$$

7.3.3. *The biological importance of γ and π in membranes*

A typical biological cell may be likened to a drop of liquid encased in a more compact liquid film, but in contrast to a drop of water the encompassing film is not of the same liquid – water – but also contains amphiphilic molecules possessing both hydrophobic and hydrophilic sections which markedly reduce surface tension to as much as half or more of that of water. Increasing the fluidity of the encompassing surface film, in a biological context, enables greater plasticity and inherent adaptability to changes of cell turgor and volume and prevents formation of detrimental membrane microlesions or "leaks". Another beneficial effect is enhanced mobility of membrane proteins which float like icebergs in a lipid sea.

7.3.4. *Measurement of surface tension and pressure*

In the following discussion only surface pressure and not tension will be mentioned, keeping in mind that the one is the converse of the other and that interconversion can be calculated from the above formula [2].

One of the most commonly employed methods for determination of surface pressure was developed by the U.S. scientist Langmuir who continued with and elaborated on observations earlier made by a German housewife Fräulein Pockels who over a century ago experimented with talc-covered soap bubbles and achieved experimental values of molecular area and pressure very closely approximating those obtained much later with far more sophisticated instrumentation. Preceding both Langmuir and Pockels was the American statesman Benjamin Franklin who amongst his many interests also experimented with surface tensions and monolayers and his writings include calculations of how many acres of lake surface one teaspoonful of olive oil would cover.

Langmuir, who in his time was the chief scientist of General Electric and later received the Nobel prize for his work, together with Blodgett developed a method whereby surface pressure, molecular area and monolayer "collapse point" can be ascertained (see Fig. 7.8). Langmuir-Blodgett films are thin organic films that can be organized into multilayer molecular assemblies one monolayer at a time. The films are made by first spreading the molecules onto an aqueous or water solution from a dilute solution after which the solvent evaporates, leaving the molecules floating on the surface. Compression leads to a densely packed monolayer (from which molecular area can be calculated by the Langmuir equation [3], see later) that if experimentally needed can be transferrcd to a solid substance by dipping the substrate through this air-water interface. Multilayers can be constructed by successive dipping through the outer surface. Only certain types of molecules, the amphiphiles, lend themselves to such behavior. As pointed out in Chapter 1 amphiphiles have two different end groups – one hydrophilic and one hydrophobic. When a monolayer, in the present context simulating an individual membrane leaflet, is formed on a water surface, amphiphiles naturally orient themselves with the hydrophilic portion in the water and the hydrophobic portion in the air. Current research effort utilizes Langmuir-Blodgett films for membrane studies and for specific technological and industrial goals (Roberts, 1990).

This method employs a trough in which an aqueous subphase subtends an overlaying monolayer of the amphiphile to be tested. The monolayer pressure is monitored by a Wilhelmy tensiometer which detects surface pressure changes upon advance of the movable teflon barrier which as seen in Fig. 7.8 progressively compresses the monolayer until its collapse. The graphically presented pressure profile is termed a *Langmuir-Blodgett surface pressure/molecular area isotherm*. Extrapolated values obtained from such isotherms are entered into the Langmuir formula [3] which enables a comparatively easy and accurate calculation of the molecular area of the individual molecules comprising the monolayer:

$$\mathrm{nm^2/mol} = \frac{A \times 10^{14}}{\mathrm{V \times M \times Avogadro's\ no.}} \qquad [3]$$

A in the above formula is the point where the tangent to the exponential section of the Langmuir isotherm intersects with the *y* axis (the dotted line in Fig. 7.8), *M* is the molarity of the monolayer, and *V* the volume of the liquid comprising the monolayer.

Fig. 7.9 indicates how the radius of a given molecule decreases with lateral compaction. Note that molecular configuration in the horizontal plane is hexagonal (Kjaer et al., 1987). As visually observed by synchrotron X-ray diffraction photography (Plate 7.5), with increase of compression, the closely packed molecular areas of a monolayer become progressively more dense and finally are arrayed in a hexagonal manner. Table 7.1 provides area measurements of some of the typical membrane constituents.

It should however be remembered that while the values presented in the table are usually of the molecule in its maximally compacted state, these values may change considerably with changes – decrease, or increase – of compaction as may occur under certain conditions in biological systems. A typical example seen in Table 7.1 is that of PIP_2 whose area decreases from 0.95 nm^2 to 0.75 nm^2 in the presence of a physiological concentration (10^{-5} M) of calcium. An interesting case of change of membrane molecule dimensions is in the highly curved regions of the thylakoid membranes in chloroplasts, where the PC molecules contained in the outer leaflet have an area of 0.72 nm^2

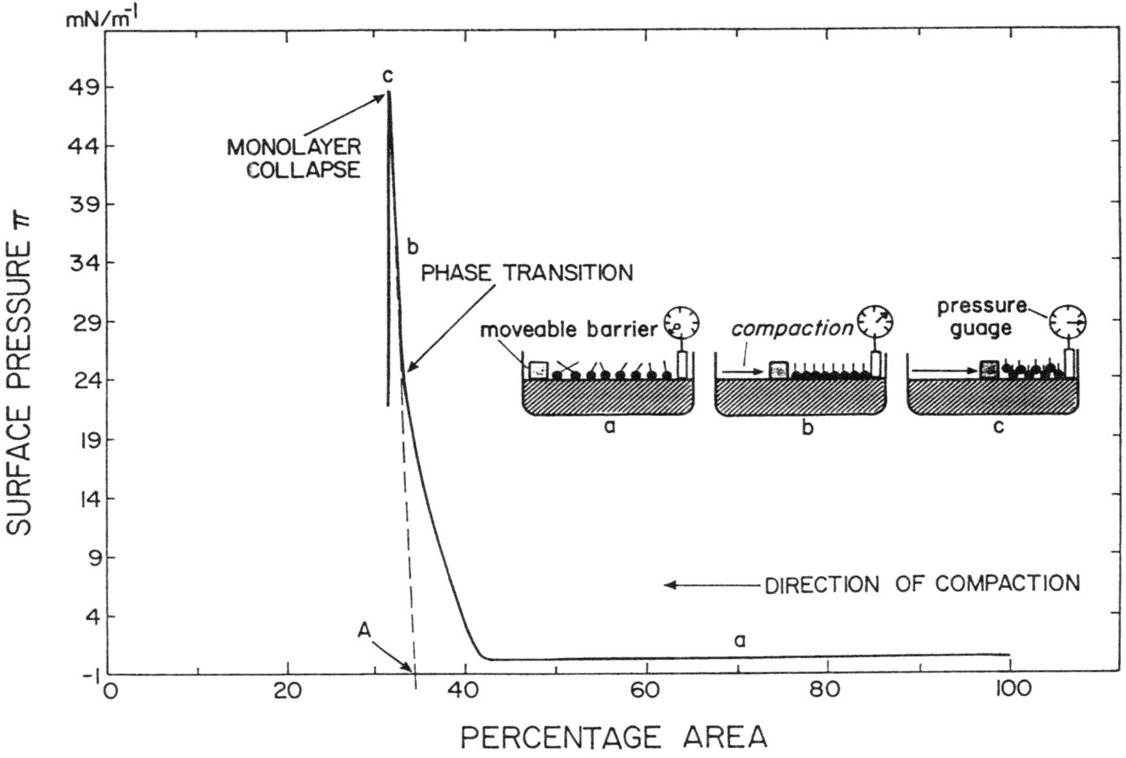

Fig. 7.8. Determination of surface pressure, molecular area and collapse point in a Langmuir-Blodgett Trough. See text for details. (In this diagram the sample employed was palmitic acid.)

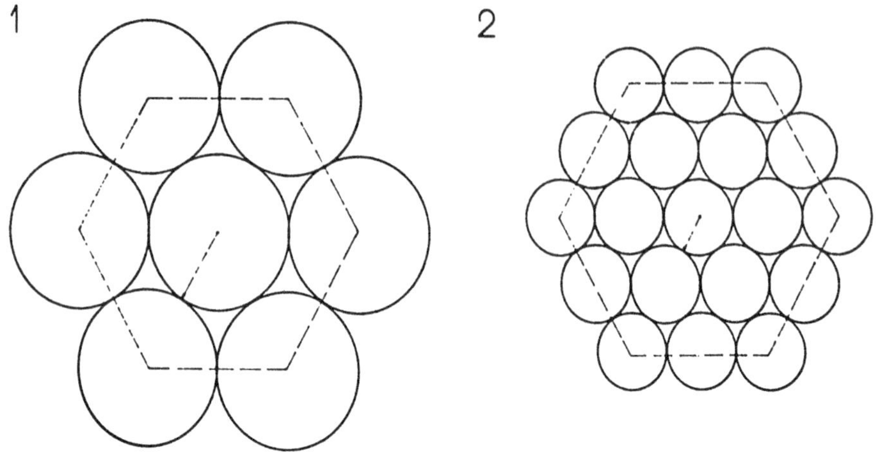

Fig. 7.9. The hexagonal lattice array of molecules in a monolayer. With increase of compression more molecules can be packed within a given area, this causing decrease of monolayer fluidity, or conversely greater rigidity.

whereas the PC molecules in the inner more highly curved leaflet possess an area of only 0.60 nm^2. Based on these and other observations the inter-relationship between molecular area, surface pressure and surface tension are presented in Fig. 7.10.

7.3.5. *Collapse point*

The *collapse point* as seen in Fig. 7.8 is the surface pressure value beyond which added compaction causes break or collapse of the monolayer. This point is actually a measure of monolayer resistance to applied pressure and as such reflects monolayer rigidity. Determination of collapse point is therefore another way to assess membrane fluidity or rigidity. In general it has been established that monolayers of glycerolipids rich in unsaturated fatty acyl tails have lower collapse points than those with more saturated fatty acyl tails. For this reason the unsaturated fatty acid tailed species are more fluid. Enhanced rigidity implies greater resistance to compaction which, while usually considered deleterious, in certain specific cases may be advantageous as in lung alveoli surfactant, stretch operated channels and stomatal guard cell membranes, more of which will be mentioned presently.

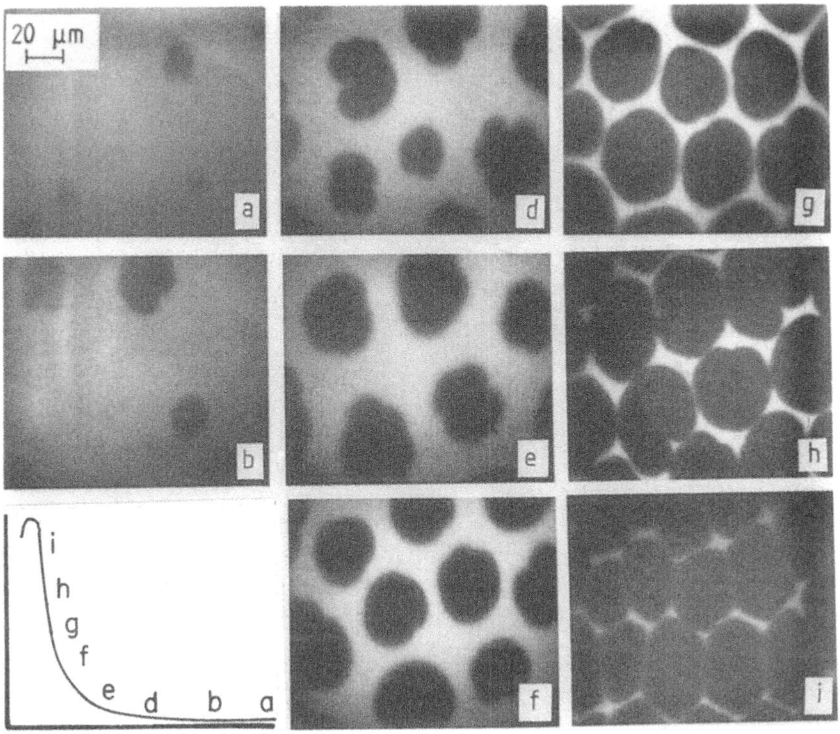

Plate 7.5. A synchrotron X-ray diffraction photograph indicating hexagonal packing in a monolayer. The letters a→i indicate points on the accompanying Langmuir isotherm at which the monolayer was photographed from above. (The plate was reproduced with the Courtesy of Prof. H. Möhwald of the University of Munich.)

Table 7.1. **Molecular areas of some amphiphiles and fatty acids present in biological membranes**

Membrane constituent	Area – nm^2	Remarks
palmitic acid	0.20	18:0 (saturated)
oleic acid	0.44	18:1 (monounsaturated)
cholesterol	0.38	–
phosphatidylethanolamine	0.41	distearic
phosphatidylcholine	0.62	*sn*-2 – oleic
phosphatidylcholine	0.69	*sn*-2 – linoleic
phosphatidylinositol-bisphosphate (PIP_2)	0.75	in presence of 10^{-5} M Ca^{2+}
phosphatidylinositol-bisphosphate (PIP_2)	0.95	no Ca^{2+}

(Data compiled from Merril and Nichols, 1985; and Landau and Leshem, 1988).

7.3.6. *Factors affecting surface tensions and pressures in membranes*

In view of the above described parameters factors affecting π and γ in the plant membrane may be arranged under the following categories:

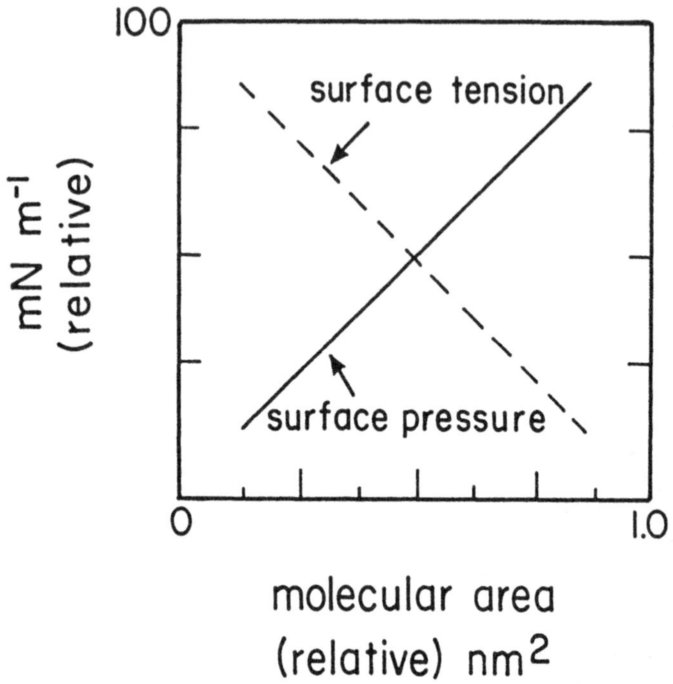

Fig. 7.10. The interrelationship between molecular area, surface tension (γ) and surface pressure (π).

a. Temperature: Earlier in this chapter it was stated that between 0°C–30°C surface tension of water varies less than 3%. Notwithstanding the behavior of water in this respect, γ and π of membranes, due to the presence of amphiphiles at the interphases, are markedly influenced in the same temperature range – the Q_{10} of a typical biological membrane in this respect being ca 1.4. This Q_{10} value is close to that of non-electrolyte diffusion in water.

b. Structural factors: Here we mention the degree of bending or conversely the "straightness" of fatty acyl chains, since membrane amphiphile molecular area depends on tail configuration as seen in Plate 2.1 and Fig. 2.3. It is quite apparent that unsaturation is directly related to fluidity. The area/molecule increases with the number of unsaturated double bonds, since a kink of ca 44° is introduced by each cis C=C bond, which implies that an increase in the degree of unsaturation of PL side-chains exerts a spacing effect between molecules in the monolayer which in turn influences the pressure behavior of the monolayer.

In contrast to headgroups, fatty acyl chains of membrane PLs have no negative charges capable of electrostatic crosslinking. This would inure these chains to pH changes. This is consistent with the observation that variation of pH between 4.5 and 7.5 caused no surface tension changes in dipalmitoyllecithin monolayers (Birdie, 1987).

Under the heading of structural factors, *polyamines* which are polycations may also be included. Their cationic charge-related effect is attributed to a membrane stabilizing effect in membrane leaflets (predominantly the inner one) which are essentially *polyanionic*. This contention also talleys with the now well-established finding that the more "poly" a polyamine is viz. the more amine groups it contains) the more physiologically effective it is. Their polycationic nature is here demonstrated by presenting the formula of spermidine which is a triamine:

$$\overset{+}{N}H_2(CH_2)_3\overset{+}{N}H(CH_2)_4\overset{+}{N}H_2$$

The following polyamines have been characterized in plants:
diamines – putrescine and cadaverine
triamines – spermidine
tetra-amines – spermine and agmatine.

The above polycation-polyamine interaction by no means excludes the possibility of polyamine action in plants in other ways which are not membrane connected. An excellent overview of plant polyamine action has been given by Galston and Kaur-Sawnhey (1983).

7.3.7. *Chemical factors*

1. *Degree of anionic charge on glycerolipid headgroups*: This aspect has been discussed in detail in Chapter 3 – here it is only rementioned that anionic charge intensity is as follows: $PIP_2 < PS < PI$ (cf. Table 3.2 on page 51). Landau and Leshem (1988) tested a series of PL monolayers PI, PS, and PIP_2 possessing an increasingly negative headgroup potential, which were spread and compressed over aqueous sub-phases and under varying conditions assayed for changes in surface parameters. The latter included surface tension and degree of molecular compression expressed by surface pressure/molecular area isotherms. Experimentation employing Wilhelmy-Du Noüy tensiometry and the Langmuir-Trough compression procedure indicated that with increasing electrostatic Ca^{2+} cross-linking, this being a direct function of Ca^{2+} concentration in the aqueous sub-phase, there was a concomitant increase in surface tension, monolayer condensation. This effect was greater with the

more negatively charged phospholipids and was more pronounced at pH 7.0 than at pH 4.5.

It can be understood that the essentially minor and negatively charged PL species may be interspersed in vivo in monolayer leaflets containing far larger amounts of electroneutral or non-anionic PLs such as PC and PEA. Hence their contribution to the net charge of the leaflet could be a major one and it is considered that the presently described model system could be applied to in vivo systems and thereby provide further information on their possible mode of action.

These negative charges may serve as sites for Ca^{2+}-mediated electrostatic cross-linking (Shah and Shulman, 1967; Morré and Bracker, 1976; Leshem and Bar-Nes, 1987). Ca^{2+} condenses monolayers of PS and PEA to a greater degree than do Na^+ and K^+; moreover, the uptake of $^{45}Ca^{2+}$ per PL molecule increases with increasing pH as determined by the number of β-particles emanating from $^{45}Ca^{2+}$-containing surfaces.

2. *pH*: This value acts in concert with "1" above and is a function both of the pH of the aqueous sections of the membrane including membrane water and cytosol, and therefore of proton pumping, and also of the anionic charge as mentioned above. An important factor contributing to the pH-mediated interplay and interdependence of the above factors is the presence of the negatively charged membrane phospholipids, in particular PS and the phosphatidylinositides – PI, PIP_2 and PIP_3 predominantly located on the inner leaflet of the membrane bilayers (Shinitzky, 1984; Batty et al., 1985; Cullis and Hope, 1986). In keeping with the view that the very acid nature of these compounds is important in cell functions, Shinitzky (1984) and van Paridon et al. (1986) have shown that below pH 5.0, PIP and PIP_2 contain 2 and 3 negative charges respectively, while at pH 8.0 they contain 3 and 5 negative charges respectively. Other contributing negatively charged glycerolipids include PS and sulpholipids (SQDG).
3. *Ca^{2+} concentration*: This factor is related to membrane rigidity – the higher the Ca^{2+} concentration the more Ca^{2+}-mediated cross-linking (cf. Fig. 5.4). Overall effects of calcium are detailed in a later chapter.
4. *Phytohormones*: Some of the phytohormones have far-reaching effects on membrane surface pressure which in certain ways may evoke similar effects as elicited by anaesthetic drugs employed in medicine. While the earliest reports relate to *ethylene*, studied together with other gases by anaesthetologists, the presently most

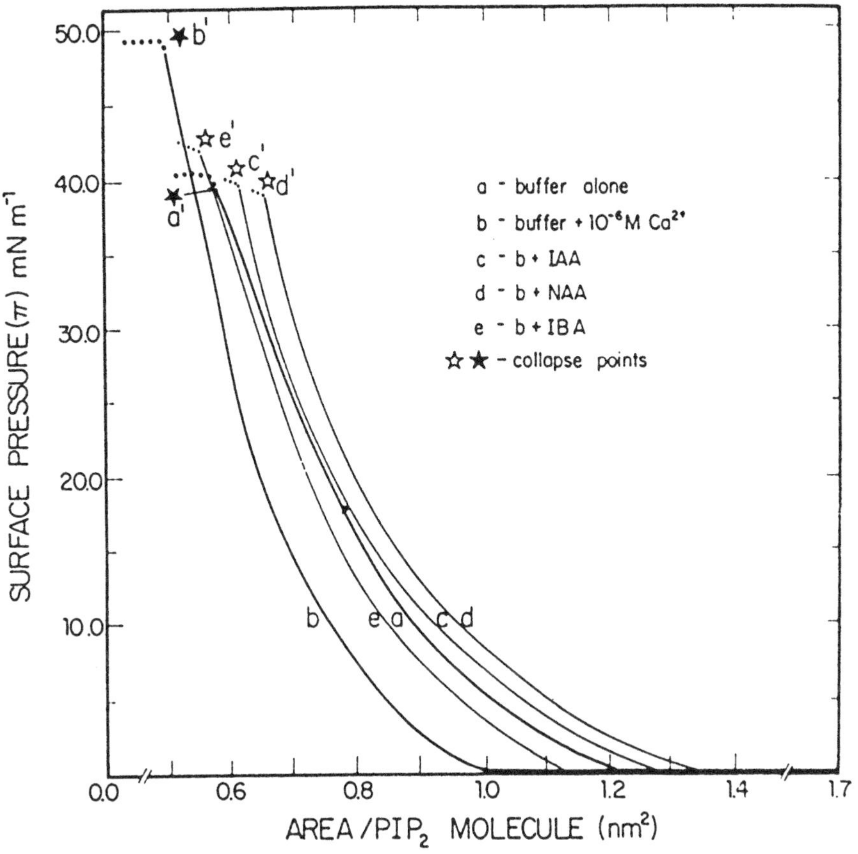

Fig. 7.11. Effect of auxins and Ca^{2+} on Langmuir molecular areal/surface pressure isotherms of a PIP_2 monolayer. Note that Ca^{2+} causes an increase of collapse point (i.e. rigidification) and a decrease in molecular area.The auxins, IAA, IBA and NAA have a converse and fluidizing effect (After Landau and Leshem, 1988).

studied plant hormone in this respect is *auxin* (indoleacetic acid) and its analogs.

Application of auxins – indoleacetic acid, indolebutyric acid and naphthaleneacetic acid (which presumably act as proton donors which may dislodge cross-linking Ca^{2+}) causes a significant decrease in surface tension when applied at physiological Ca^{2+} concentrations. As evidenced by surface pressure/molecular area isotherms, Ca^{2+} *increased* monolayer rigidification and *decreased* PIP_2 molecular area. The auxins tested possessed an opposite and derigidifying effect as indicated by increase of the phospholipid's molecular area and by the observed considerable lowering of monolayer collapse pressures (Fig. 7.11). The surfactive effectivity of the auxins increased with decreasing sub-phase Ca^{2+} concentration

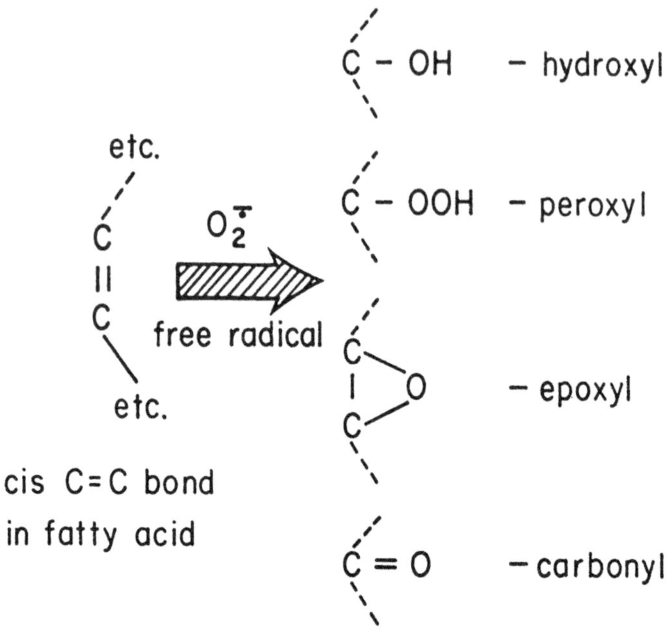

Fig . 7.12. Fatty acyl peroxidation products resulting from free radical attack (After Bamberger et al., 1988).

in the 10^{-4} to 10^{-6} mol dm^{-3} range. These data collectively suggest that surface-associated biophysical changes should be taken into account during interpretation of transmembrane hormonal signal transduction, the efficiency of which is impaired during senescence.

It has been suggested that auxin binds to membrane phospholipid headgroups (Jones and Paleg, 1984), and in this way may affect surface tension. A further assumption concerning the putative mode of this hormone's effect is that being an acid it may supply protons which could dislodge Ca^{2+} from electrostatic bridges (Landau and Leshem, 1988).

5. *Free radicals*: The unsaturated C═C bonds in membrane fatty acyl sidechains are particularly sensitive to free radical attack especially under conditions which promote overall organ senescence. As a result of free radical action several compounds in which the cis 44° "kinked" section of the chain undergoes a chemical change (see Fig. 7.12) which results in straightening-out of the chain. When this occurs *less* molecules occupy a given surface area of membrane and rigidification ensues – this change also involves deceleration of diffusional movements of embedded proteins. Free radicals may also impair membranes in other ways, an extensive discussion of which has been presented in the volume by Leshem et al. (1986).

6. *Sterols*: Sterol effects on membrane rigidity and surface tension have been discussed in Chapter 4 (qv).

7.4. Brownian and diffusional movements in membranes

In hypothesizing on the nature of movement of suspended molecules of submicroscopic size in liquids, Einstein (1956) has stated that classical thermodynamics cannot be looked upon as applicable with precision. Such molecules include amphiphilic membrane glycerolipids. Visual observation of the above mentioned submicroscopic particles was originally made by the botanist Robert Brown in 1828 who noted that movement of pollen grains suspended in a drop of water was "uninterrupted and swarming". Such Brownian movement was ascribed by Gouy in 1888 to the effect of the thermal molecular motions of the suspending liquid. Formulae enabling calculation of these motions which were found to be both *rotational* and *lateral* (or as common designated, *translational*) were worked out by Einstein. With certain modifications (Saffman and Delbruck, 1975) his work is still regarded as a basis for behavior of amphiphiles and similar molecules in liquids. Fig. 7.13 depicts these diffusional movements. In biomembranes the two diffusional components are:

a. *Rotational diffusion* – designated D_r which is a measure of the number of revolutions a molecule performs around itself in a given unit of time.
b. *Lateral (or translational) diffusion* – designated D_T or sometimes D_l which is the mean square of the distance per unit time that a molecule moves laterally. It is the constant of proportionality relating the mass transport per unit area to the concentration gradient.

Rotational diffusion takes place around the longitudinal axis of glycerolipids and, according to Rupert (1988), also has a slight wobble, – 28° to each side of the perpendicular to the molecule's axis (see Fig. 7.13). Thus in a given plane in a membrane leaflet the constituent molecules both gyrate and move laterally. Phospholipids in a liquid crystalline state (Lα) diffuse rapidly in the order of 1–4×10^{-8} cm^2/sec – this value being considerably decreased when phospholipid aggregates are formed.

Diffusional movement of membrane proteins is markedly affected by frictional forces exerted by their lipid milieu and generally is 10–10,000 × slower than lipid diffusidon. The anisotropic structure of the membrane bilayer allows D_r to occur along an axis roughly *perpendicular to the bilayer's plane*, while almost completely excluding such

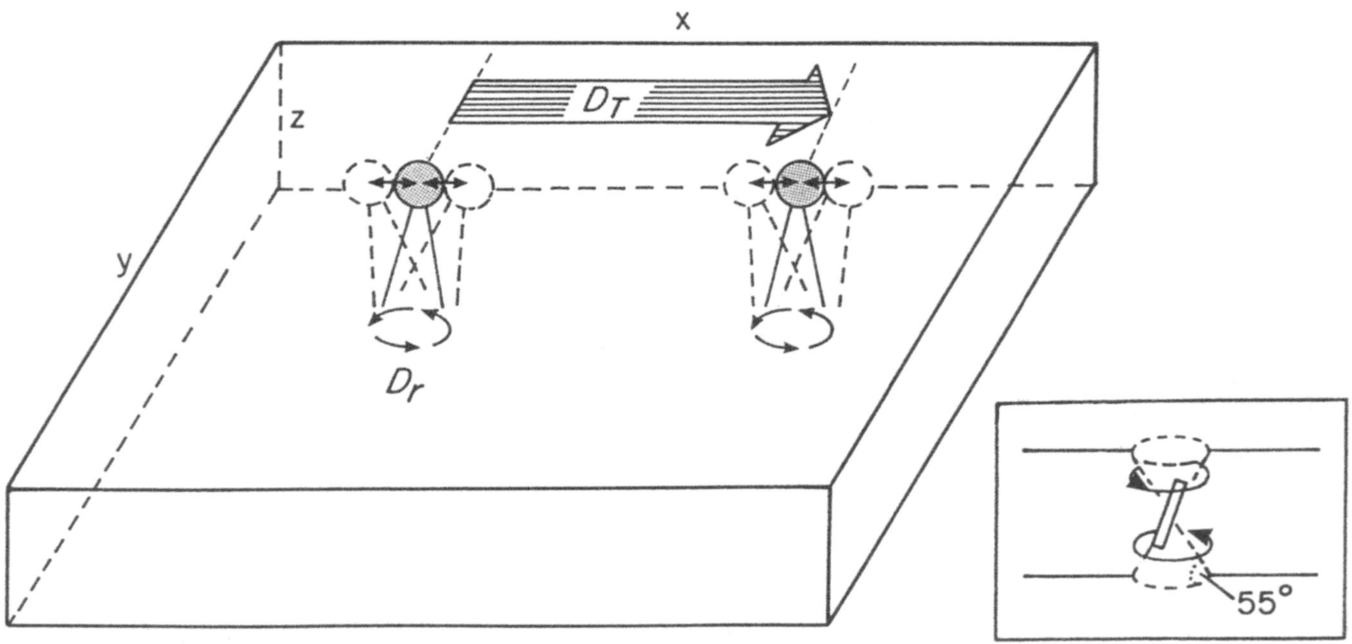

Fig. 7.13. Types of diffusion in a biological membrane. D_r is *rotational diffusion* and is a measure of the angular velocity of the molecule rotating around its axis. This would affect monolayer surface tension by friction of either PL-PL molecules or PL-H_2O molecules in the aqueous interstices. D_r is given by the Einstein formula (see equation [4]). Note that a certain degree of wobble ca 28°) exists. D_T is *translational* or *lateral diffusion* and is the mean square of the distance per unit time that a molecule moves laterally. It is the constant of proportionality relating the mass transport per unit area to the concentration gradient (see equation [5]).
This diagram conceals the fact that statistically-calculated *average* tilt of the revolving molecules is 55° (i.e., ca 1 radian) and not vertical. This is seen in the *insert*.

rotational movement along the plane parallel to the bilayer (Shinitzky, 1984; Merril and Nichols, 1985). In other words membrane proteins spin like tops rather than like gyroscopes. For the same reason, lateral diffusion occurs along a plane *parallel to the membrane plane* (see Fig. 3.13).

Calculations, the details of which will be presented later, have shown that an individual phospholipid molecule in an unbound state can change places with its neighbor 10^7 times per sec. and possesses a lateral diffusional constant in the order of 4×10^{-8} cm^2/sec. The following scheme on an ascending scale indicates the motion frequencies of some of the various membrane components:

Motion	Frequency (sec^{-1})
lipid flip flop	10^{-2}–10^{-6} (days, weeks)
sterol flip flop	↓
protein rotational diffusion	
protein lateral diffusion	
lipid lateral diffusion	
lipid rotational diffusion	10^8–10^{10} (nanosecs, picosecs)

As stated above D_T values of proteins are slower. In soybean (*Glycine max*) protoplasts protein D_T values varying from 3×10^{-9} cm^2/sec (fast) to 5×10^{-10} cm^2/sec (slow) have been reported (Metcalf et al., 1980, 1986) – the slower rate being attributed either to the presence in the membrane of large scale immiscible gel and fluid domains or to a certain degree of hindrance caused by anchorage to cytoskeletal elements. In contrast, in rose petal protoplast's cytoskeleton disrupting drugs had little effect on protein diffusion whereas elevation of $CaCl_2$ concentration in the medium from 1 mM to 10 mM reduced the mobile fraction of membrane proteins (Walko and Nothnagel, 1989). Wier and Edidin (1988) have presented an overall and numerical categorization of translational diffusion coefficients of membranes and have included them in three ranges:

i) $D_T > \sim 3 \times 10^{-9}$ cm^2/sec. Diffusion of these proteins is largely constrained by the viscosity of membrane bilayer lipids.

ii) $D_T < 1 \times 10^{-12}$ cm^2/sec. This range is characteristic of proteins that are immobile on the time scale of measuring equipment. Such proteins often form patches or clusters in the membrane. Translational diffusion of these proteins is probably limited by their interaction with sub-membranous cytoskeleton and by self-aggregation. The acetylcholine receptor is a typical example.

iii) A diffusion range intermediary between i and ii this being typical of most membrane proteins. Cytoskeletal elements seem to place constraint on this group.

7.4.1. *Calculation of D_r and D_T*

In the first decade of this century Albert Einstein provided the foundations for understanding and calculation of Brownian and diffusional movements. Readers interested in his original papers translated into English are referred to the annotated booklet – *Investigations on the Theory of the Brownian Movement* – published in 1956. Following are the Einstein equations:

The Einstein diffusional formulae:

$$D_r = \frac{K_B T}{8\pi\eta r^3} \frac{1}{\text{sec}} \qquad [4]$$

rotational diffusion

$$D_T = \frac{K_B T}{6\pi\eta r} \text{cm}^2/\text{sec} \qquad [5]$$

translational or lateral diffusion

In these equations K_B is the Boltzmann constant, T the absolute temperature, r the measured molecular radius, and η the microviscosity in poise units. Note that the denominator of Eq. [4] is the friction coefficient.

Strictly speaking the above formulae only apply to a spherical molecule, whereas glycerolipids are of irregular shape. However, since an exact analytic calculation of the friction coefficient of these irregularly shaped molecules is virtually impossible, the friction coefficient as employed in Eq. [4] can be regarded as an effective approximation and may serve as a basis for comparison.

With the unravelling of molecular shapes and dimensions of membrane fatty acids as achieved by de Jong (see Fig. 2.3 in Chapter 2) it is now possible to enter the r (molecular radius) values in the above formulae: such calculation also employing reported areas of membrane phospholipids (Merril and Nichols, 1985) provide a means of establishing diffusional coefficients. As an example, Table 7.2 presents D_T and D_r values of typical unsaturated fatty acids found in plant membranes.

Table 7.2. D_T and D_r values of typical membrane-contained cis-unsaturated fatty acids

No. of unsaturated C=C bonds	Molecular radius (nm)	D_r	D_T
1	0.57	100	100
2	0.88	32	69
3	0.94	22	61

(After Merril and Nichols, 1985; and Leshem, 1989). The fatty acids are 18:1, 18:2 and 18:3. Values are represented on a relative scale.

This table indicates an inverse relationship between degree of unsaturation and radius on the one hand, and D_r and D_T on the other.

However as stated by Einstein himself the above formulae only apply to an ideal situation in which *spherical* molecules are in an *isotropic* plane: thus any biological calculations basing on these formulae are only approximations. Saffman and Delbruck 1975 of the University of California have modified Einstein's formulae for anisotropic situations as could be encountered in biological membranes. They found that in contrast to the isotropic situation, anisotropicity causes a fourfold increase of D_T as compared to D_r. However here again calculations were based if not on spherical molecules, then on geometrical cone-shaped ones and not on irregularly structured biological components.

A somewhat more up-to-date development of translational diffusional calculations, which pays especial attention to changes that could result from lipid peroxidation has been presented by Richter (1987). But he also bases his calculations on a conical model while fully cognizant that no accurate calculation can be made of the highly non-geometric glycerolipid or protein configurations. Richter's amended formula for D_T takes the viscosity of fluid housing the diffusing molecule into account:

$$D_T = (K_B T/4\pi\eta h)[\ln(\eta h/\eta_\omega a) - \gamma] \qquad [6]$$

In this formula D_T is translational diffusion; K_B is the Boltzman constant; T – the absolute temperature; η – the microviscosity of fluid phased membrane; η_ω – the microviscosity of the medium; γ – Euler's constant; h – the height of the molecule; a – molecular radius, this being analogous to r in the Einstein formulae.

7.4.2. *Experimental determination of D_T*

Several techniques are available to measure in vivo lateral diffusion in membranes, however *fluorescence recovery after photobleaching* (FRAP), the development of which was pioneered by Michael Edidin of the Johns Hopkins University, is by far the most widely used one. In this technique a uniformly fluorescent probe-labeled membrane is bleached in one spot using a laser beam. The enhanced fluorescence from the illuminated spot is monitored as a function of time after bleaching, and the rate of "recovery" (viz. entry of non-bleached molecules into the laser treated spot via lateral diffusion) is a direct measure of D_T of the fluorescent species from the surrounding medium into the previously illuminated spot. Most of the plant and mammalian research on D_T has employed FRAP, which despite the doubts as to what, other than bleaching, the laser beam could do to the cell components – this technique is regarded as less damaging than others. A list of some of the commonly used fluorescent probes is provided by Gennis (1989).

7.5. Piezobiophysics of the membrane

In biophysical terminology the prefix *piezo-* is used instead of the adjectival phrase *pressure-related*, and this branch of piezobiophysics deals with the effects of pressure on biological systems. A typical piezophysical effect is known to most readers who, when driving an automobile, apply pedal pressure to the braking system – this pressure being hydraulically transferred to the brake lining on the wheels via the non-compressible braking fluid filling the connecting rigid metal tubing. The property of *compressibility* (c) plays an important role in the physiological adaptation of living species (piscian and mammalian, including whales and man) to conditions of high pressure and also according to a certain school of thought may be employed to counteract pain. In membrane function the compressibility may be affected by membrane constituent – each group possessing different c values – and hence physiological responses may be regulated. An extensive review on piezophysics has been presented by Heremans (1982). Table 7.3 give c values of some compounds pertinent to our subject matter.

This table indicates that the cell's cytosol – which is essentially an aqueous system containing proteins – possesses very low compressibility. In contrast the encompassing membrane bilayer, providing

Table 7.3. Comparative compressibilities (c*) at 25°C of some biological compounds (After Shinitzky)

Compound or system	Compressibility (atm. 10^{-5})
proteins	1.0
water	4.7
hexane	17.0
membrane bilayer	
liquid-crystalline phase	24.0
solid-gel phase	0.0

$$*c = \left[\frac{1}{V}\left(\frac{\delta V}{\delta P}\right)\right] \times 10^{-5}\ \text{atm.}$$

that it is in a liquid crystalline state, has a considerably higher compressibility value. It thus transpires that when a biological tissue is exposed to pressure changes, the cell is equipped with an effective means of adaptation. The table also suggests that within the bilayer the most effective component is the phospholipids since the proteins and the water are virtually non-compressible. Upon transition to the solid-gel phase, membrane compressibility is completely lost, as resultingly could be the ability to withstand pressure changes.

As outlined earlier in this book the non-compressibility of proteins has been used to develop a "gentle" hydraulic pressure-based method of membrane isolation which yields biologically and immunologically active proteins (Fig. 5.12). Leshem and Bar-Nes (1987) have adapted this method to plant membranes and experiments performed on pea foliage microsomal membranes show that presence of Ca^{2+} in the medium decreases amount of protein release while auxin increases it.

From all stated above it follows that biological tissues exposed to repeated cycles of atmospheric or hydrostatic compaction and decompaction, e.g. lung alveoli, stomatal guard cells etc. should be, and indeed are, equipped with a means to enable them to withstand piezobiophysical changes. Without these, under excessive pressure cells would burst, and under extremely low pressures, collapse. In lung alveoli cell rigidity – like a pneumatic tire protecting an inner tube – is obtained by the presence in *lung surfactant* of a phosphatidylcholine species in which *both* fatty acyl tails are saturated – this specific PC species being *dipalmitoylphosphatidylcholine* (DPPC). The DPPC occurs as a monolayer at the air:liquid interface of pulmonary tissue and is found both in the intra- and extracellular lung surfactant (Harwood, 1987; Post and van Golde, 1988).

In the author's laboratory DPPC has been detected in plant stomatal guard cell membranes albeit in low amounts and in these membranes the analogy is to enforcing rigid threads which are present in a relatively small proportion of an essentially pliable matrix as can be found for example in nylon-enforced cotton textile cloth or steel-threaded pneumatic tires (Leshem et al., 1990). The opening and closing of "stretch operated channels" mentioned in Chapter 5 are probably also subjected to piezobiophysical regulation.

In conclusion to the present chapter dealing with membrane biophysics I wish to point out that the surface pressure/tension parameter experimental approach is fundamentally different from that based upon assays of microviscosity, phases, compressibility, etc. The difference lies in that in the latter, membranes isolated from biological tissues are employed, while in the former the experimentation is only simulatory. The simulation data is indeed exact but it is obtained only from assay of membrane constituents and not on in vivo biological tissue, and the analogies drawn from the detail to the general are inferred to reflect pertinent aspects of membrane and possibly of whole-organism behavior. The biologically-obtained data may apparently provide more reliable information, but has the drawback that possibilities of exact pinpointing of mechanisms is limited: hence the ideal combination of both approaches.

7.6. References and recommended reading

Aloia, R.C. (Ed.). 1983–5. *Membrane Fluidity in Biology. Vols. I–III.* Academic Press, New York, London.

Bamberger, E.S., Alter, M., Landau, E.M. and Leshem, Y.Y. 1988. Biophysical effects of superoxide on surface parameters of a model membrane. pp. 113–6. In: Eds. O. Hayaishi, C. Niki, M. Kondo and T. Yoshikawa. *The Medical, Biochemical and Chemical Aspects of Free Radicals.* Elsevier, Amsterdam.

Batty, I., Nahorski, S.R. and Irvine, R.F. 1985. Rapid formation of inositol 1,3,4,5 tetrakisphosphate following muscarinic receptor stimulation of rat cerebral cortical slices. *Biochem. J.* 232:211–5.

Birdie, K.S. 1987. Dipalmitoyllecithin monolayer at the air/water interface. *Langmuir* 3:132–3.

Borochov, A., Halevy, A.H., Borochov, H., and Shinitzky, M. 1978. Microviscosity of rose petals' plasmalemma as affected by age and environmental factors. *Plant Physiol.* 61:812–5.

Brenner, R.R. 1984. Effect of unsaturated fatty acids on membrane structure and enzyme kinetics. *Prog. Lip. Res.* 23:69–96.

Brown, D.J., Dupont, F.M. 1989. Lipid composition of plasma membranes prepared from roots of barley. *Plant Physiol.* 90:955–61.

Coolbear, K.P., Berde, C.B. and Keough, K.M.W. 1983. Gel to liquid-crystalline phase transitions of aqueous dispersions of polyunsaturated mixed acid phosphatidylcholines. *Biochemistry* 22:1466–73.

Crowe, J.H., Crowe, L.M., Carpenter, J.F. and Winston, C.H. 1987. Stabilization of dry phospholipid bilayers by sugars. *Biochem. J.* 242:1–10.

Crowe, J.H., Hoekstra, F.A. and Crowe, L.M. 1989. Membrane phase transitions are responsible for imbibitional damage in dry pollen. *Proc. Natl. Acad. Sci. USA* 86:520–3.

Cullis, P. and Hope, M.J. 1986. Physical properties and functional roles of lipids in membranes. pp. 25–72. In: Eds. D.E. Vance and J.E. Vance. *Biochemistry of Lipids and Membranes*. Benjamin Cummings, Menlo Park, California.

Davis, P.J., Fleming, B.D., Coolbear, K.P. and Keough, K.M.V. 1981. Gel to liquid-crystalline transition temperatures of water dispensions of two pairs of positional isomers of unsaturated mixed acid phosphatidylcholines. *Biochemistry* 20:3633–6.

Deckman, M., Haimovitz, R. and Shinitzky, M. 1985. Selective release of integral proteins from human erythrocytes by hydrostatic pressure. *Biochem. Biophys Acta* 821:334–40.

de Gier, J., van Echleld, C., van der Steen, A., Noordau, P.C., Verklij, A.J. and de Kruijff, B. 1982. Lipid organization and barrier functions of membranes. pp. 315–25. In: Eds. J. Wintermans and P. Kuipers. *Biochemistry and Metabolism of Plant Lipids*. Elsevier, Amsterdam.

de Kruijff, B., Cullis, P.R. and Verklij, A.J. 1980. Non bilayer lipid structures in model and biological membranes. *TIBS* 79–81.

Einstein, A. 1956. *Investigations on the Theory of the Brownian Movement*. Ed. with notes R. Furth. Dover Press, New York.

Galston, A.W. and Kaur-Sawnhey, R. 1983. Polyamines: are they a new class of plant growth regulators? pp. 451–96. In: Ed. P.E. Wareing. *Plant Growth Substances 1982*. Academic Press, New York.

Gennis, R.B. 1989. *Biomembranes: Molecular Structure and Function*. p. 184. Springer Verlag, Berlin-Heidelberg.

Harwood, J.L. 1987. Lung surfactant. *Prog. Lip. Res.* 26:211–56.

Heremans, K. 1982. High pressure effects in proteins and other biomolecules. *Ann. Rev. Biophys. Bioeng.* 11:1–21.

Jones, C.P. and Paleg, L.G. 1984. Complex formation between IAA and phospholipid components in aqueous media. 2. Interaction of auxins and related compounds with phosphatidyl membranes. *Biochemistry* 23:1521–4.

Kjaer, K., Als-Nielsen, J., Helm, C.A., Laxhuber, L.A.A. and Möhwald, H. 1987. Ordering in lipid monolayers. *Phys. Rev. Lett.* 58:2224–6.

Landau, E.M. and Leshem, Y.Y. 1988. Biophysical interactions of membrane anionic phospholipids with pH, calcium and auxins. *Journ. Exp. Bot.* 39:1689–97.

Larsson, C., Møller, I.M. and Widell, S. 1990. Introduction to the plant plasma membrane – its molecular composition and organization. In: Eds. C. Larsson and I.M. Møller, pp. 1–15. *The Plant Plasma Membrane*. Springer Verlag, Berlin-Heidelberg.

Legge, R.L., Cheng, K.H., Lepock, J.R. and Thompson, J.E. 1985. Differential effects of senescence on the molecular organization of membrane in ripening tomato fruit. *Jour. Physiol.* 81:954–9.

Lentz, B. 1989. Membrane fluidity as detected by diphenylhexatriene probes. *Chem. Phys. Lipids* 50:171–90.

Leshem, Y.Y. 1987. Ca^{2+} and intermolecular bridging of membranal phospholipids and proteins. pp. 225–7. In: Eds. P.K. Stumpf, J.B. Mudd, W.D. Ness. *The Metabolism, Structure and Function of Plant Lipids*. Plenum Press, California.

Leshem, Y.Y. 1989. The function of membrane glycerolipid fatty acid unsaturation and cell cycle control: a biophysical approach. pp. 198–202. In: Eds. P. Biacs, T.

Gluiz and T. Brenner. *The Biological Role of Plant Lipids*. Plenum, New York and Academiai Kiado, Budapest.

Leshem, Y.Y. 1991. Evidence for the presence and mode of action of a membrane associated plant phospholipase A_2. pp. 53–5. In: Eds. P.L. Quinn and J.L. Harwood, *Plant Lipid Biochemistry. Structure and Utilization*. Portland Press, Colchester and London.

Leshem, Y.Y. and Bar-Nes, G. 1987. Hormone receptor manipulation by hydrostatic pressure: interaction between calcium: membrane components and PI in pea foliage membranes. pp. 155–62. In: Ed. D. Klämbt. *Plant Hormone Receptors*. Springer Verlag, Berlin.

Leshem, Y.Y., Cojocaru, M., Magel, S., El-Ani, D. and Landau, E.M. 1990. A biophysical study of abscisic acid interaction with membrane phospholipid components. *New Phytol.* 116:487–98.

Leshem, Y.Y., Halevy, A.H. and Frenkel, C. 1986. *Processes and Control of Plant Senescence*. Elsevier, Amsterdam. pp. 61–3.

Leshem, Y.Y. and Inbar, M. 1978. Resistance to gibberellin induced changes of lipid fluidity in wheat embryo mitochondrial membranes as assessed by the fluorescent probe, DPH. *J. Exp. Bot.* 29:671–5.

Leshem, Y.Y., Landau, E.M. and Deutsch, M. 1988. A monolayer model study of surface tension-associated parameters of membrane phospholipids: effect of unsaturation of fatty acyl tails. *Jour. Exp. Bot.* 39:1679–87.

Leshem, Y.Y., Sridhara, S. and Thompson, J.E. 1984. Involvement of calcium and calmodulin in membrane deterioration during senescence of pea foliage. *Plant Physiol.* 75:329–35.

Lindblom, G., Brentał, T., Sjolund, M., Wikander, G. and Weslander, A. 1986. Phase equilibria of membrane lipids from *Acholeplasma laidlawii*: Importance of a single lipid forming non-lamellar phases. *Biochem.* 25:7502–10.

Lindblom, G. and Rilfors, L. 1989. Cubic phases and isotropic structures formed by membrane lipids – possible biological relevance. *Biochim. Biophys. Act.* 988:221–56.

Lindstedt, S. and Liljenberg, C. 1990. On the periodic minimal surface structure of the plant prolamellar structure. *Physiol. Plant* 80:1–4.

McKersie, B., Thompson, T.E. and Brandon, J.K. 1976. X ray diffraction evidence for decreased lipid fluidity in senescent membranes from cotyledons. *Can. J. Bot.* 54:1974–8.

Merril, A.H. and Nichols, J.W. 1985. Techniques for studying phospholipid membranes. pp. 61–96. In: Ed. J.F. Kuo. *Phospholipids and Cellular Regulation, Vol. II*. CRC Press, Boca Raton.

Metcalf, T.N., Villanueva, M.A., Schindler, M. and Wang, J.L. 1986. Monoclonal antibodies directed against protoplasts of soybean cells: analysis of the lateral mobility of plasma-membrane antibody MVS-1. *Jour. Cell. Biol.* 102:1350–7.

Metcalf, T.N., Wang, J.L. and Schindler, M. 1980. Lateral diffusion of phospholipids in the plasma membrane of soybean protoplasts: evidence for membrane lipid domains. *Proc. Nat. Acad. Sci. USA* 83:95–9.

Morré, D.J. and Bracker, C.E. 1976. Ultrastructural alteration of plant plasma membrane induced by auxin and calcium ions. *Plant Physiol.* 58:544–7.

Morré, D.J., Crowe, J.H., Morré, D.M., and Crowe, L.M. 1987. Infrared spectroscopic evidence for a conformational alteration of plant plasma membranes upon exposure to the growth hormone analogue, 2,4-D. *Biochem. Biophys. Res. Comm.* 147:506–12.

Nur, T., Ben Arie, R., Lurie, S. and Altman, A. 1986. Involvement of divalent cations in maintaining cell membrane integrity in stressed apple fruit tissue. *Jour. Pl. Phys.* 125:47–60.

Palmgren, M.G. and Sommarin, M. 1989. Lysophosphatidylcholine stimulates ATP dependent proton accumulation in isolated oat root plasma membrane vesicles. *Plant Physiol.* 90:1009–14.

Pauls, K.P. and Thompson, J.E. 1980. In vitro simulation of senescence related membrane damage by ozone induced lipid peroxidation. *Nature* 283:504–9.

Pauls, K.P. and Thompson, J.E. 1982. Effects of cytokinins and antioxidants on the susceptibility of membranes to ozone damage. *Plant Cell Physiol.* 23:821–32.

Pike, C.S. 1982. Membrane lipid physical properties in annuals grown under contrasting thermal regimes. *Plant Physiol.* 52:236–9.

Platt-Aloia, K.A. and Thompson, W.W. 1987. Freeze fracture evidence for lateral phase separations in the plasmalemma of chilling injured avocado fruit. *Protoplasma* 136:71–80.

Post, N. and van Golde, L.M.G. 1988. Metabolic and developmental aspects of the pulmonary surfactant system. *Biochim. Biophys. Act.* 947:249–86.

Quinn, P. 1989. The phase behaviour of membrane lipids and the organization of the photosynthetic membrane. pp. 209–15. In: Eds. P. Biacs, K. Gruiz and T. Kremmer. *Biological Role of Plant Lipids*. Akad. Kiado, Budapest. Plenum Press, New York.

Raison, J.K., Pike, C.S. and Berry, J.N. 1982. Growth temperature induced alterations in the thermotropic properties of *Nerium oleander* membrane lipids. *Plant Physiol.* 70:215–8.

Rawyler, A. and Siegenthaler, P.A. 1989. Changes in the molecular organization of MGDG between resting and functional states of thylakoid membranes. pp. 225–6. In: Eds. P. Biacs, K. Gruiz and T. Kremmer. *Biological Role of Plant Lipids*. Akad. Kiado Budapest. Plenum Press, New York.

Richter, C. 1987. Biophysical consequences of lipid peroxidation in membranes. *Chem. Pys. Lip.* 44:175–90.

Roberts, G. (Ed.) 1990. *Langmuir-Blodgett Films*. Plenum, New York.

Rojas, E. and Tobias, J.M., 1965. Membrane model association of inorganic cations with phospholipid monolayers. *Biochim. Biophys. Acta*, 94:394–404.

Rupert, L.A.M. 1988. The head group conformation and dynamical behavior of didodecylphosphate bilayer membranes. *Jour. Coll. Interface Sci.* 125:286–94.

Saffman, P.C. and Delbruck, M. 1975. Brownian movement in biological membranes. *Proc. Nat. Acad. Sci. USA* 72:3111–3.

Schroeder, F. 1984. Role of membrane lipid asymmetry in aging. *Neurobiol. Aging* 5:323–33.

Selstam, E., Brentel, J. and Lindblom, G. 1990. The phase structure of galactolipids and their role in the formation of the prolamellar body. 14:843–8. In: Ed. M. Balcheffsky. *Current Research in Photosynthesis III*. Kluwer Academic Publishers, Dordrecht, The Netherlands.

Shah, O.H. and Schulman, J.H. 1967. The ionic structure of lecithin monolayers. *Jour. Lipid Res.* 8:227–33.

Shinitzky, M. 1984. Membrane fluidity and cellular functions. pp. 1–15. In: Ed. M. Shinitzky. *Physiology of Membrane Fluidity, Vol. 1*. CRC Press, Boca Raton, Florida.

Shinitzky, M. and Barenholz, Y. 1978. Fluidity parameters of lipid regions determined by fluorescence polarization. *Biochem. Biophys. Acta* 515:367–94.

Shinitzky, M. and Inbar, M. 1974. Difference in microviscosity induced by different cholesterol levels in surface membrane lipid bilayer of normal lymphocytes and malignant lymphoma cells. *Jour. Mol. Biol.* 85:603–15.

Siegel, D.P. 1986. Inverted intermediates and the transitions between lamellar, cubic and inverted hexagonal amphiphile phases. *Chem. Phys. Lip.* 42:279–301.

Sklar, L.A. 1980. The partition of *cis*-parinaric acid and *trans*-parinaric acid among aqueous fluid lipid and solid lipid phases. *Molec. Cell. Biochem.* 32:169–75.

Skyamsander, E., Gruner, S.M., Tate, M.W., Turner, D.C. and So, P.T.C. 1988. Observation of inverted cubic phase in hydrated dioleoylphosphatidylethanolamine membranes. *Biochemistry* 27:2332–6.

Sridhara, S. and Leshem, Y. 1986. Phospholipid catabolism and senescence of pea foliage membranes: parameters of Ca^{2+}: calmodulin: phospholipase A_2 induced changes. *New Phytol.* 102:5–16.

Stillwell, W., Brengle, B., Belcher, D. and Wassal, S.R. 1987. Comparison of effects of ABA and IAA on phospholipid bilayers. *Phytochem.* 26:3145–50.

van der Meer, W. 1984. Physical aspects of membrane fluidity. pp. 53–71. In: Ed. M. Shinitzky. *Physiology of Membrane Fluidity, Vol. 1.* CRC Press, Boca Raton, Florida.

van Paridon, P.A., de Kruijff, B., Ouwerkerk, R. and Wirtz, K.W.A. 1986. Polyphosphoinositides undergo charge neutralization in the physiolgical pH range; a ^{31}P NMR study. *Biochim. Biophys. Acta* 877:216–9.

Vigh, L., Horvath, I., Woltjes, T., Farkas, P., van Hasselt, P.J.C. and Kuiper, P. 1987. Combined ESR, X-ray diffraction studies on phospholipid vesicles obtained from cold hardened wheats. *Planta* 170:14–9.

Walko, R.M. and Nothnagel, E.D. 1989. Lateral diffusion of proteins and lipids in the plasma membrane of rose protoplast. *Protoplasma* 152:46–56.

Wier, M. and Edidin, M. 1988. Constraint of the translation diffusion of a membrane glycoprotein by its external domains. *Science* 242:412–4.

Wills, R.H., Lee, T.H., Graham, D., McGlasson, W.B. and Hall, E.G. 1981. Postharvest. *An Introduction to the Physiology and Handling of Fruit and Vegetables.* Avi Publishers, Westport, CT.

Part three

The membrane in action

8. The role of calcium in the membrane

Research in the recent decade has clearly indicated that Ca^{2+} has graduated from the exclusive club of elements which partake in plant mineral nutrition and has become a full-fledged, or even a major factor whose importance in biological regulation, according to some scientists, has elevated it to the rank of a "secondary messenger". (The term should not be confused with mRNA etc. which belongs to an entirely different field of biological regulation.) This chapter makes no claim to deal with Ca^{2+} metabolism in its entirety and will only deal with its structural and functional roles which are related to the topic of plant membrane biophysics. Detailed treatment of Ca^{2+} in overall plant physiology extending beyond the membrane realm include the thorough reviews of Hepler and Wayne (1985) and of Allan and Trewavas (1987).

The aspects to be discussed are as presently follow: in general the order of presentation is that the material provided at each stage serves as a basis for understanding of the successive one:

a. The chemistry of Ca^{2+}.
b. Structural and biophysical aspects of presence of Ca^{2+} in the mem-rane.
c. Membrane calcium distribution, gradients and interaction with calmodulin.
d. Applied aspects.
e. Ca^{2+} and membrane fusion.

8.1. The chemistry of Ca^{2+}

A glance at the Periodic Table of the elements (cf Table 8.1) shows that Ca^{2+} and several other elements belong to Group IIa, which comprise the "alkaline earth metals".

Table 8.1. Group IIa of the Periodic Table of elements – the alkaline earth metals

PERIOD	GROUP IIa
2	Be
3	Mg
4	Ca
5	Sr
6	Ba
7	Ra

While all the elements belonging to this group are divalent, and at least theoretically could substitute for each other in biological processes, it is well known that in plant tissues calcium and magnesium are the most prevalent. Indeed these two elements are interchangeable in various processes which presently will be described; there are however marked chemical differences between them which usually cause the Ca^{2+} to be preferred over the Mg^{2+}. Since the other elements in this group are of rare occurrence in plants our discussion will therefore center around Ca^{2+} and Mg^{2+}, which are the most prevalent in mammals as well.

All elements belonging to the IIa Periodic Table group may be likened to localized and spherical charges in which the electrostatic fields are *directly* related to the electric charges and are *inversely* related to the squares of the ionic radii. Ca^{2+} has an ionic radius of 99 picometers (pm) while that of Mg^{2+} is approximately *one third less*, being only 65 pm. Because of the latter's smaller radius it binds to H_2O with greater attractive forces than Ca^{2+}, as for example do colloidal clay particles as compared to larger soil particles such as loam or sand. Resultingly, more force is necessary to remove H_2O from the magnesium than from the calcium ion. Fig. 8.1 compares the sub-atomic structures of these two elements and depicts the basic difference between them regarding their respective electron orbital and suborbital shells. Pauling has pointed out that the outer electrons (nos. 1, 2 and 3) in the alkaline earth metal group are readily removable and thus may form the appropriate cations – Ca^{2+}, Mg^{2+}, etc. Fig. 8.1 moreover shows that in contrast to Mg^{2+}, Ca^{2+} possesses a "*d*" orbital and it is believed that this allows the formation of a greater

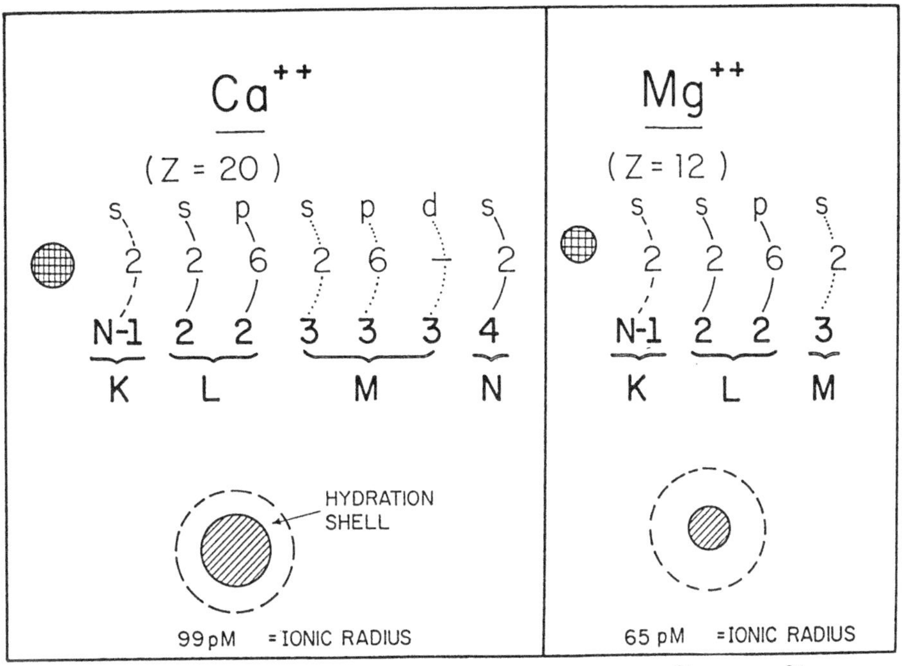

Fig. 8.1. Comparison of the sub-atomic structures of Ca^{2+} and Mg^{2+}.

number of complexes having co-ordination numbers (from 6–12) of varying lengths.

In summary, it appears that due to Ca^{2+}'s greater ionic radius and its potential ability to form bonds of varying lengths, it tends to fit into clefts or atomic niches in or between molecules which serve as membrane building-blocks far more readily than Mg^{2+}. Hepler and Wayne (1985) have assessed that the affinity for such binding of membrane components to Ca^{2+} is 1000 times more than the affinity for Mg^{2+}.

8.2. Structural and biophysical aspects of presence of Ca^{2+} in the membrane

Ca^{2+} may constitute a structural element in membrane architecture, and at low concentrations functions as a stabilizer, while at high ones, as a rigidifying agent. These two functions stem from its divalent nature and also, as outlined in section "a" above, from its ionic radius. Both properties enable Ca^{2+} to form electrostatic bridges either between individual neighboring anionic phospholipids, or between the latter and membranes, or with cytoskeletal elements. These possibilities have been dealt with in more detail on pages 70–72 in Chapter 5 (see also Fig. 5.4).

Besides effects on membrane fluidity, Ca^{2+} exerts a clearcut effect on membrane surface tension-related parameters: when brought into contact with physiological concentrations of Ca^{2+}, atomic radii of membrane glycerolipids become much smaller, and concomitantly their compressibility (and hence rigidity) increase. It is of interest that the auxin group of phytohormones may reverse the compression induced by calcium, and this hormonal mechanism should also be taken into account when interpreting the basic mode of action of auxins (Landau and Leshem, 1988). The explanation offered is that the protons which are liberated by the auxin (indolacetic acid) may partially remove the cross-linking Ca^{2+} from the electrostatic bridges which rigidify the membrane.

It is improbable that in a membrane all the anionic sites would bind with Ca^{2+} or with some other divalent cation. If this indeed were so the membrane would become exceedingly rigid, protein (enzyme or receptor) movement would be sluggish and due to high surface tension that would be formed, translocatory function of membranes would be minimal. On the other hand were it not for partial Ca^{2+}-mediated cross-linking, membrane domains would become excessively fluid and hence unstable; under certain conditions embedded proteins may jut out or sink into the bilayer too deeply for optimal activity which requires optimal stereochemical and topographical configurations for protein/receptor/substrate recognition. In extreme cases hyperfluidity could even lead to the complete ejection of embedded proteins from the bilayer.

Another facet of Ca^{2+} presence in membranes is in connection with the activity of some of the integral membrane proteins such as Ca^{2+}ATPase. This will be discussed in the next section.

8.3. Membrane calcium distribution, gradients and interaction with calmodulin

From an ever-increasing number of researches in animal and plant systems it has now become evident that a Ca^{2+} gradient of several orders of magnitude exists between the outer to inner side of the bilayer. The cytosolic intracellular Ca^{2+} concentration is within the range 10^{-6}–10^{-7} M while the extrafacial membrane region has a Ca^{2+} content in the 10^{-3} M range (Roux and Slocum, 1982). It thus transpires that the cell's cytosolic interior is relatively poor in Ca^{2+} whereas

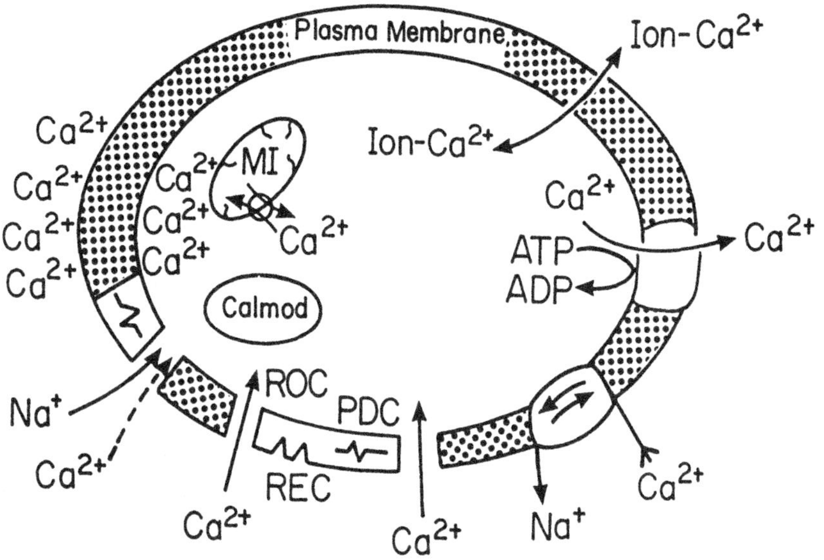

Fig. 8.2. Controlling mechanisms of cellular calcium movements. Calcium influx can occur through ion channels, including the fast sodium channel (a minor component) and via the potential dependent (PDC) and receptor operated (ROC) calcium channels. Additionally, calcium influx can occur as one component of a plasmalemmal sodium-calcium antiporter exchange process that can operate in either direction. Calcium pumping is represented by a plasmalemmal calcium-ATPase, but calcium sequestration (and release) can take place at several intracellular sites, including mitochondria (MI), the endoplasmatic reticulum (ER), the vacuole, chloroplasts and the internal surface of the plasma membrane. A non-physiologic calcium entry process is represented by ionophore transport (ion-Ca^{2+}). Calmod = calmodulin. (After Triggle et al., 1982).

the main and potentially biologically active store of Ca^{2+} is on the membrane's exterior. It must however be stressed that secondary stores of high Ca^{2+} content exist in cellular organelles such as mitochondria, in the endoplasmatic reticulum and apparently also in the chloroplast and here again gradients of several orders of magnitude exist between the endofacial and exofacial sides of the membrane. Other stores of calcium, which as opposed to the above are generally considered biologically inert, are the Ca-oxalate crystals in the vacuole and the Ca-pectate in the middle lamellae intervening between proximal cells.

According to current concepts the translocation of even small amounts of Ca^{2+} from one side of the membrane to the other – either inwards or outwards – serves as a trigger to several basic regulatory processes and hence the concept of "*Ca^{2+} as a secondary messenger.*" Fig 8.2 summarizes possibilities of Ca^{2+} passage through biological membrane bilayers. Optimal cellular function depends upon maintaining the above-described outer/inner Ca^{2+} gradient; when the cell senesces or when cell integrity is impaired by pathogenic or mechanical means, leaks are formed, massive unregulated and danger-fraught entry of external Ca^{2+} into the cytosol ensues.

8.3.1. *Ca^{2+} influx into the cytosol*

Ca^{2+} influx in plants as in mammals can take place through receptor operated calcium channels, the receptor creating electric potential between the two sides of the channel (Dieter, 1984). A further path of entry into the plasmalemma is a bidirectional $Na^{+}:Ca^{2+}$ antiport exchange. Another possible non-physiological entry of Ca^{2+} is mediated by either endogenous or exogenously applied Ca^{2+} *ionophores*. These ionophores cause no lysis of membrane structure but cancel the Ca^{2+} gradient and allow Ca^{2+} concentrations on either of the membrane to equilibrate according to kinetics of regular diffusional flow. A typical Ca^{2+} ionophore that is a handy and widely used tool in research is A23187 which is a lipophilic antibiotic produced by *Streptomyces chartreusensis* (Kauffman et al., 1980). Increasing the concentration of intracellular free Ca^{2+} with ionophore A23187 more than doubled the number of androgenic pollen grains in *Solanum carolinense* and could partially substitute for 2,4-D in inducing pollen callus formation (Reynolds, 1990). Another such ionophore which is regarded as being more specific than A23187 is *ionomycin* which also is an antibiotic and is produced by a different micro-organism – *Streptomyces conglobatus* (Albert and Tashjian, 1984).

These ionophores are not absolute and also have ionophoretic effect, albeit weak, on other divalent cations. The order of selectivity is as follows:

$$Ca^{2+} > Mg^{2+} > Sr^{2+} = Ba^{2+}$$

As mentioned in the chapter dealing with membrane fatty acyl tail-groups, a contention – which still awaits experimental affirmation – exists whereby peroxidized polyunsaturated fatty acids produced by the action of the enzyme lipoxygenase, as well as phosphatidic acid may serve as specific Ca^{2+} ionophores. It may well be that these catabolites cause enhanced Ca^{2+} efflux or influx, but this may not necessarily be a classic ionophore response since the fluxes may stem from membrane microlesions which they could cause, and so doing equilibrate Ca^{2+} concentrations on either side of the bilayer.

Ca^{2+} influx may be forestalled not only by pumping in the opposite direction (detailed in next section) but also by application of *Ca^{2+} channel blockers*. These blockers were developed by the pharmaceutical industry with the aim of preventing the undesirable loss of Ca^{2+} from various tissues, and some of these are of widespread medicinal use. Ca^{2+}-orientated plant research has been aided by application of these specific Ca^{2+} channel blockers which include verapamil, nitrenedipine, dialtizem, cannarizine, flunarizine and chloropomazine as well as lanthanum chloride ($LaCl_3$). Experimental use of these blockers indeed has prevented Ca^{2+} accumulation in plant cells and at the same time has promoted plant growth (Hetherington and Trewavas, 1984; Vaughan and Mulkey, 1985). At present the applied agricultural use of Ca^{2+} channel blockers is not feasible since the possibility exists of human consumption of blocker-treated produce in which case non-desirable side effects may be caused.

8.3.2. *Ca^{2+} efflux form the cytosol*

As can be understood from the previous section it stands to reason that Ca^{2+} efflux must take place *against* an osmotic gradient. This phenomenon may be encountered upon exit of calcium into the intercellular solution or upon entry of Ca^{2+} to cytosol-contained organelles. Efflux mechanisms may be grouped under two headings:

i) Active pumping by the Ca^{2+} ATPase group of enzymes: In plants enzymes of this category have been found in the plasmalemma, in mitochondrial membranes and to a certain extent also in chloroplast membranes. The activity of the Ca^{2+}ATPase appears to be dependent upon the structure of the phospholipids surrounding it in the membrane. Medium-length fatty acyl chains promote optimal activity since this enzyme reconstituted into bilayers of dipalmitoyl (18 C) PC is more active than that of the enzyme reconstituted into either dimyristoyl (14 C) or, ditetracosanoic (24 C) PC (Michelangeli et al., 1991).
ii) Activated diffusion through channels across which a localized electric potential exist, these differing from the overall $\Delta\psi$ of the membrane and allowing Ca^{2+} translocation against the gradient.

Concerning mitochondria Dieter (1984) and others have demonstrated the importance as activators, of NAD kinase and of calmodulin in respiratory processes. Hence the consensus that in addition to trans-membrane Ca^{2+} movement through the plasmalemma which apparently accounts for the major bulk of this cation's translocation, the efflux from the cytosol into these two organelles cannot be ignored.

However, in the wake of research conducted in animal systems, it has been demonstrated that the Ca^{2+} movement in the mitochondrion is possibly from compartment to compartment *within* the mitochondrion and not from the encompassing cytosol and the organelle (Denton and McCormack, 1985). In plants, to the best of our knowledge, this point has as yet not been clarified conclusively. We shall therefore suffice with the conclusion that indeed Ca^{2+} translocation regulates NAD kinase activity, and that at present it cannot be stated from where to whence. A more detailed discussion of this issue can be found in a minireview on the role of Ca^{2+} in plant membrane senescence (Leshem, 1987). Other targets of Ca^{2+} fluxing are the endoplasmatic reticulum and the vacuole. The latter was formerly regarded as an inert sub-cellular organelle serving as a "garbage-dump" for discard products: in view of its contribution to Ca^{2+} concentration regulation and the now well established catabolite recycling that occurs in the organelle, the garbage dump concept should be reappraised and modified. In this connection a Ca^{2+}/H^+ antiporter located in the tonoplast appears to function in the release of vacuole-stored calcium (Schumaker and Sze, 1984; 1990). Tonoplast-located Na^+/H^+ and K^+/H^+ antiporters were also discussed in Chapter 5 and the relevance to osmo-regulation and resistance to salt stress there discussed.

8.3.3. *Calmodulin and its antagonists*

Calmodulin: (etymologically derived from *calcium modulator*) is a tetrameric Ca^{2+} binding protein discovered in the 70's by the U.S. scientist G. Cheung. Depending on source, calmodulin is 16.7–18.0 kD protein – a strong resemblance existing between bacterial, plant and mammalian types thus suggesting a marked evolutionary conservation of Ca^{2+} binding mechanisms. Regarding cellular and subcellular distribution, an opposite situation to that of Ca^{2+} is encountered. In plant cells it is mainly located in the cytosol (ca 90%), very little is found in the membrane, and it is virtually absent from the outer membrane space (Muto, 1982). Collinge and Trewavas (1984) have found that calmodulin comprises approximately 3% of total plasmalemmal protein in peas. Smaller secondary stores of calmodulin are found associated with the nucleus.

Fig. 8.3 shows its mode of action. As further demonstrated in Fig. 8.4 calmodulin, which in itself is biologically inactive, upon binding to Ca^{2+} forms an active calmodulin:Ca^{2+} complex. Calmodulin thus serves as a Ca^{2+} sensor which is activated when the concentration of this divalent cation increases in response to physiological excitation.

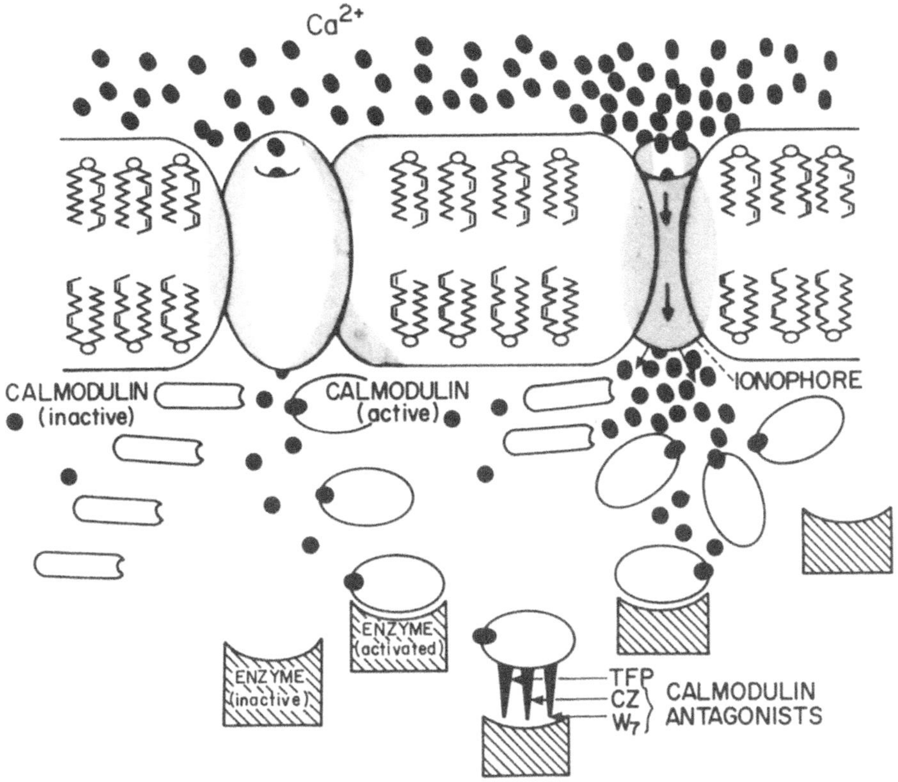

Fig. 8.3. Inter-relationships between Ca^{2+}, calmodulin and Ca^{2+}:calmodulin sensitive enzymes. tfp = trifluoperazine; cz = calmidozolium.

The Ca^{2+} causes a stereochemical change in calmodulin structure and resultingly the complex is able to bind to specific sites present in a group of Ca^{2+}:calmodulin-sensitive enzymes (see Fig. 8.5). Some of these enzymes have an obligatory requirement for the Ca^{2+}:calmodulin complex; the action of others taking place without calmodulin, is considerably enhanced when it is included in addition to Ca^{2+}.

Ca antagonists: Calmodulin research has been greatly aided by use of *phenothiazines* which are a group of calmodulin antagonists, their pharmacological use being in the field of clinical psychiatry where they are administered as anti-depressants and in the treatment of schizophrenia. A commonly employed phenothiazine in calmodulin research is *trifluoperazine* (TFP). As said these compounds are anti-calmodulin affectors – however it has now become evident that this antagonistic effect on calmodulin is limited to low concentrations of

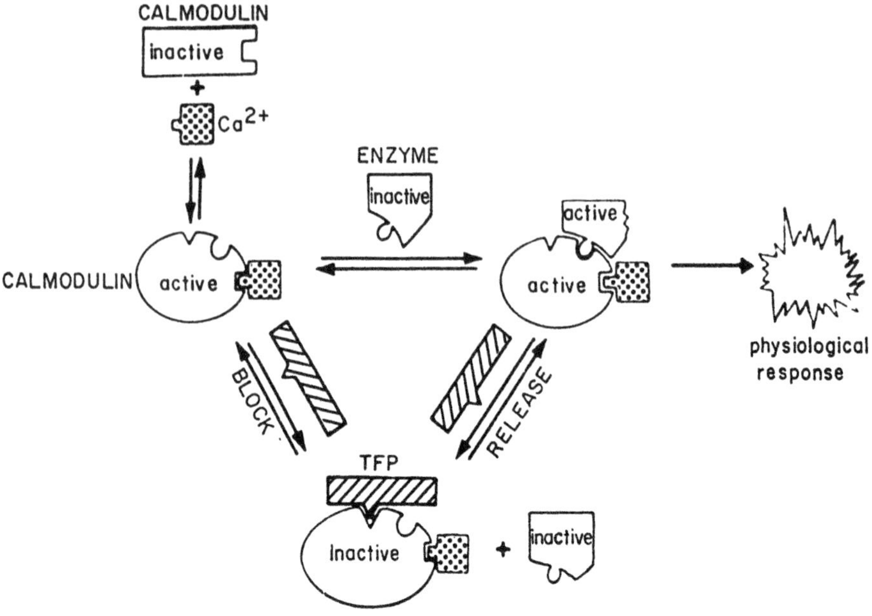

Fig. 8.4. Enzymatic regulation by a Ca^{2+}:calmodulin complex and mode of inhibition by psychotic drugs such as trifluoperazin. (Adapted from Weiss and Wallace, 1980).

these psychotic drugs, higher ones having been reported to elicit non-specific membrane effects.

A second group of calmodulin antagonists also serving as local anaesthetics are compounds such as *β-endorphin* and *calmidazolium* (=R24571). Japanese research has now produced a series of so-called *W-compounds* of which W-7 -(N-aminohexyl)-5 chloro-1-naphtalene sulphonamide hydrochloride) is claimed to be the most specific (Eliotet et al., 1983). Research carried out in the author's laboratory and also in collaboration with research in Thompson's laboratory in Waterloo, Canada has shown that the Ca^{2+}:calmodulin complex on the one hand promotes pea foliage senescence, while on the other hand the calmodulin antagonists TFP and W-7, serving to forestall Ca^{2+} complexing with calmodulin, have a senescence-inhibiting effect (Leshem et al., 1984; Sridhara and Leshem, 1986).

Another process where Ca^{2+}:calmodulin interaction seems to play a regulatory role is stomatal movement. De Silva et al. (1985) and Schwartz et al. (1988) have suggested that the unhindered movement of the stomate-closing hormone, abscisic acid, is dependent upon the free trans-membrane Ca^{2+} movement. Upon foliage treatment with the Ca^{2+} antagonists TFP and W-7, abscisic acid action was blocked. The issue of Ca^{2+} in stomatal movement has been reviewed by Mansfield et al. (1990). Pertinent to the calmodulin issue and signal trans-

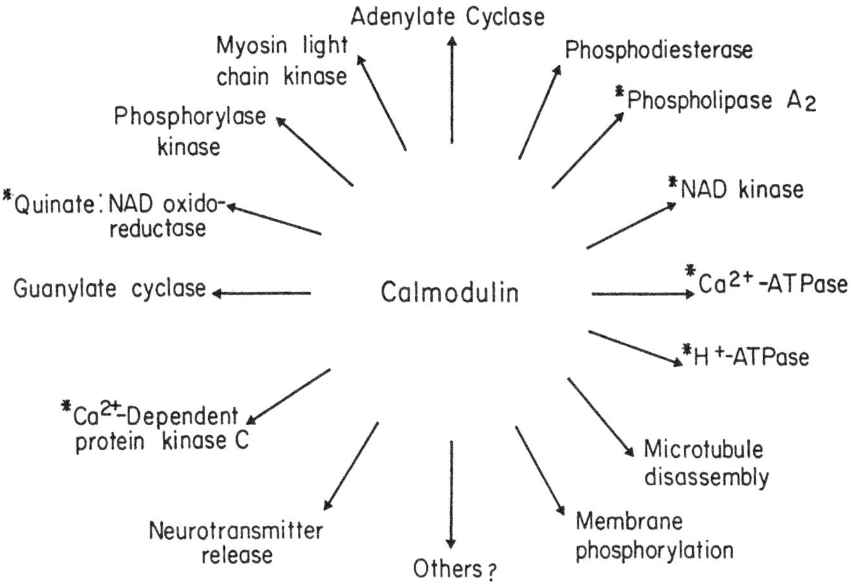

Fig. 8.5. Enzymes depending on Ca^{2+}:calmodulin complexing for activation. Asterisks indicate those identified in higher plants.

duction are the reports of Ca^{2+} and calmodulin dependent protein kinase C in some plant species while others contain a protein kinase which is Ca^{2+}-dependent but calmodulin-independent (Harmon et al., 1987).

A further effect of Ca^{2+} is in photosynthetic sucrose synthesis and it has been suggested that in this respect Ca^{2+} may be acting at three sites. The first is at the level of *fructose* 1, 6 *bisphophatase* which would be limited if Ca^{2+} increased; a second potential target is *sucrose phosphate synthase* activation which appears to be both Ca^{2+} and calmodulin dependent. The third target is surmised to be *inorganic pyrophosphate* turnover since Ca^{2+} inhibits the tonoplast pyrophosphate-dependent protein pump (Brauer et al., 1990).

Before reading the following section which deals with practical and applied aspects of Ca^{2+} interaction with membranes, the reader is referred to Chapter 12 dealing with "signal transduction". Perusal of the said chapter will add to the better understanding of what is stated herewith.

8.4. Applied and agricultural aspects of membrane Ca^{2+} metabolism

In previous chapters we have seen that Ca^{2+} that does not leak into the cytosol may act as a membrane stabilizer. Based upon this and upon early observations of Poovaiah and Leopold in 1973, a method has been developed whereby deciduous fruit, primarily apples, are treated with $CaCl_2$ which is administered to fruit tissue by pressurization. The principle is placing post-harvest apples in a 4% $CaCl_2$ solution in a pressure drum: the calcium solution enters the fleshy tissue through scar tissue located at the distal part of the fruit where the flower style was situated or via microlesions on fruit epidermi and cuticules. According to the experience of Ferguson's group in New Zealand, accumulation of Ca^{2+} is in the intercellular spaces – viz. the physiologically termed *free space* – and is not internalized in the *osmotic volume*. This pressure treatment with the calcium solution increases the Ca^{2+} gradient between the outside and the inside of fruit cell membranes and so doing stabilizes and enables them to better withstand onset of senescence processes. Fruit treated in this manner has a far longer "shelf-life" and may be shipped to longer distances while preserving freshness. Moreover the physiological disease, *brown spot*, which is manifested by the appearance of brown discolored tissue in the white flesh especially just under the peel, appears significantly later or not at all in Ca^{2+}-treated apples. An alternative to pressure treatment with $CaCl_2$ is $Ca(NO_3)_2$ spray applied several weeks before fruit picking – results from spray application in terms of post-harvest physiology are similar to those obtained with pressurization (Ferguson and Watkins, 1981).

Ca^{2+} also counteracts stress associated phenomena in fruit, an effect attributed to membrane stabilizing action (Nur et al., 1986; Lurie et al., 1987; Ferguson and Drøback, 1988; Glenn et al., 1988; Stow, 1989).

In the above described cases the common denominator is an inhibition or deceleration of senescence. One of the manifestations of the latter is fruit maturation which goes hand-in-hand with development of fruit aroma and fragrance; subsequently Ca^{2+}-treated produce is sometimes of a somewhat inferior quality with regards to taste and aroma. However the economic advantages of producing fruit better equipped against marketing-associated deterioration and loss usually override other considerations.

A further means of prevention of a too early onset of senescence is the blocking of Ca^{2+} entry into the cytosol which ordinarily would trigger Ca^{2+}-requiring phospholipase action either with or without Ca^{2+}. As explained in the chapter on signal transduction this involves the occurrence of a series of reactions in which *inter alia* free radicals, ethylene, malondealdehyde etc. are formed – all of which are known to promote cell catabolism and ageing. Use of Ca^{2+} channel blockers prevents the initial Ca^{2+} entry stage and at least in an experimental system have increased production of roots in maize (Vaugham and Mulkey, 1985). As yet because of potential side-effect complications, use of channel blockers is limited to experimental systems.

A further avenue of interest in Ca^{2+}-related membrane physiology is being pursued in Halevy's laboratory at the Faculty of Agriculture of the Hebrew University in Rehovoth. His group's research indicates that Ca^{2+} participates in the photoperiodic response of certain ornamental plants especially in the "*light break*" phenomenon. In the "light break" reaction a brief pulse of light given to plants in the middle of the dark period under short-day conditions causes them to react as if they were in long days, with the resulting effect on flower induction or inhibition. It appears that translocation of Ca^{2+} through the membranes is an important factor in this light break induced flowering control (Friedman et al., 1989). This conclusion tallies with the earlier contention of Raven (1983) that both phototropic and photoperiodic responses in plants are intricately connected with the ability of membranes to allow passage of the Ca^{2+} ion. In this context we mention the intriguing observation of Roux and Slocum (1982), which has subsequently been substantiated, that *phytochrome* when converting from the P_{660} to the P_{730} form (red to far red) induces physiologically important changes in the orientation of membrane components thus allowing facilitated Ca^{2+} translocation.

To summarize the issue of Ca^{2+} effects on membrane-associated plant senescence in brief – it may be generalized that when located on the membrane's exterior Ca^{2+} promotes, and when on the interior it inhibits, senescence. Chapter 12 will delve deeper into some of the purported biochemical and biophysical changes associated with Ca^{2+} entry into the cytosol.

8.5. Ca^{2+} and membrane fusion

Membrane fusion in biological processes often is faced with a difficulty stemming either from the presence of membrane proteins (since initially the lipid sections of the bilayer leaflets fuse) or from the binding of membrane components with cytoskeletal elements which anchor membrane and make fusion cumbersome, if not impossible. Working on human erythrocytes, Murachi and co-workers in Japan unravelled the existence of a protease with an obligatory requirement for Ca^{2+} which they termed *calpain*. By severing the bonds between membrane components and sub-bilayer cytoskeletal proteins, calpain

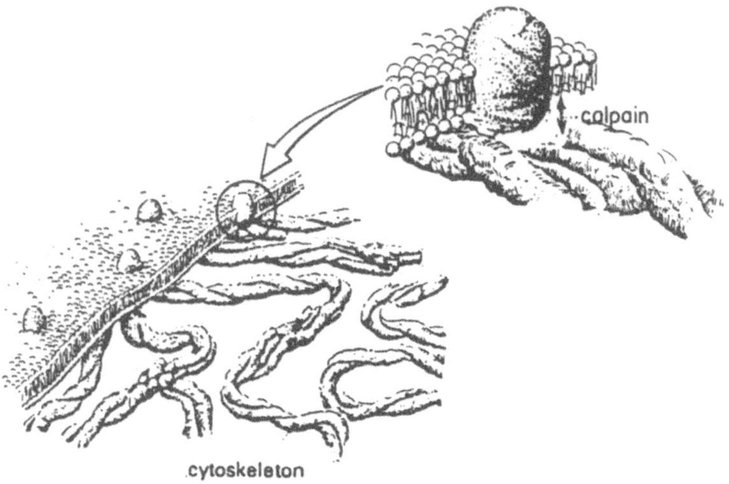

Fig. 8.6. Ca^{2+} activated protease, calpain. Calpain may facilitate membrane fusion by severing contact between the bilayer and cytoskeletal elements. Another enzyme, calstatin, has an opposite and stabilizing effect.

releases the constraints imposed by the cytoskeleton on membrane fusion (see Fig. 8.6). Later work by Murachi and colleagues in Japan, by Kosower at the Medical School of the Tel-Aviv University, and by others, have moreover revealed the existence of another calcium dependent enzyme – *calstatin* – which has the opposite effect to calpain in that it stabilizes membrane:cytoskeletal-bonding against calpain action (Murachi et al., 1981; Kosower et al., 1983; Mellgren, 1987).

While this topic to the best of our knowledge has not yet been investigated in the Plant Kingdom, in plant biotechnology where "protoplast" fusion between two populations with different DNA contents often produces low percentages of recombinants which amongst other factors, is due to incomplete membrane fusion, the addition of calpain would be a worthwhile endeavor. This is all the more so since calpain extracted from erythrocytes has been found to be active in other non-erythrocyte membrane categories and appears not to be organ or species specific.

8.6. References and recommended reading

Albert, P.R. and Tashjian, A.H. 1984. Relationship of thyrotropin releasing hormone induced spike and plateau phases in cytosolic Ca^{2+} concentrations to hormone secretion. *Jour. Biol. Chem.* 259:15350–63.

Allan, E.F. and Trewavas, A.J. 1987. The role of calcium in metabolic control. pp.

117–49. In: Ed. P.D. Davies. *The Biochemistry of Plants, Vol. 2. Physiology of Metabolism*. Acad. Press, New York.

Brauer, M., Sanders, D. and Stitt, M. 1990. Regulation of photosynthetic sucrose synthesis: a role for calcium. *Planta* 182:236–43.

Collinge, M. and Trewavas, A.J. 1984. The location of calmodulin in the pea plasma membrane. *Jour. Biol. Chem.* 264:8865–72.

Denton, R.M. and McCormack, J.C. 1985. Physiological role of calcium transport by mitochondria. *Nature* 315:634.

de Silva, D.L.R., Cox, R.C., Hetherington, A.M. and Mansfield, T.A. 1985. Suggested involvement of calcium and calmodulin in the responses of stomata to abscisic acid. *New Phytol.* 101:555–63.

Dieter, P. 1984. Calmodulin and calmodulin mediated processes in plants. *Plant, Cell Environ.* 7:371–80.

Eliot, D.C., Batchelor, S.M., Cosoor, R.A. and Mosinos, N.C. 1983. Calmodulin binding drugs affect responses to cytokinim, auxin and gibberellic acid. *Pl. Phys.* 40:175–80.

Ferguson, I.B., Drøback, B.K. 1988. Calcium and the regulation of plant growth and senescence. *Hort. Sci.* 23:262–6.

Ferguson, I.B. and Watkins, C.B. 1981. Ion relationds of apple fruit tissue during fruit development and ripening II. Calcium uptake. *Austral. Jour. Plant Phys.* 8:259–66.

Friedman, H., Goldschmidt, E.E. and Halevy, A.H. 1989. Involvement of calcium in the photoperiodic flower induction process of *Pharbitis nil*. *Plant Phys.* 89:530–4.

Glenn, G.M., Reddy, A.S.N. and Poovaiah, B.W. 1988. Effect of calcium on cell wall structure, protein phosphorylation and protein profile in senescing apples. *Plant Cell Phys.* 29:565–72.

Harmon, A.C., Putnam-Evans, A. and Cormier, M.J. 1987. A calcium dependent but calmodulin independent protein kinase from soybean. *Plant Physiol.* 83:830–7.

Hepler, P.K. and Wayne R. 1985. Calcium and plant development. *Ann. Rev. Pl. Phys.* 36:397–439.

Hetherington, A.M. and Trewavas, A.J. 1984. Binding of nitrenedipine, a calcium channel blocker to pea shoot membranes. *Pl Sci. Lett.* 35:109–13.

Kauffman, R.K., Taylor, R.W. and Pfeiffer, D.R. 1980. Cation specificity of ionomycin comparison with ionophore A23187 in rat liver mitochondria. *Jour. Biol. Chem.* 235:2735–9.

Kosower, N.S., Glaser, T. and Kosower, E. 1983. Membrane mobility agent promoted fusion of erythrocytes: fusibility is correlated with attack by calcium activated cytoplasmic proteases on membrane proteins. *Proc. Nat. Acad. Sci. U.S.A.* 80:7542–6.

Landau, E.M. and Leshem, Y.Y. 1988. Biophysical interaction of membrane anionic phospholipids with pH, calcium and auxins. *Jour. Expt. Bot.* 39:1689–97.

Leshem, Y. 1987. Membrane phospholipid catabolism and Ca^{2+} activity in control of senescence. *Physiol. Plant.* 69:551–9.

Leshem, Y.Y., Halevy, A.H. and Frenkel, C. 1986. *Processes and Control of Plant Senescence*. Elsevier Press, Amsterdam.

Leshem, Y., Sridhara, S. and Thompson, J.E. 1984. Involvement of calcium and calmodulin in membrane deterioration during senescence of pea foliage. *Pl. Phys.* 75:329–35.

Lurie, S., Sonego, L. and Ben Arie, R. 1987. Permeability, microviscosity and chemical changes in the plasma membrane during storage of apple fruit. *Scient. Hort.* 32:73–83.

Mansfield, T.A., Hetherington, A.M. and Atkinson. 1990. Some current aspects of stomatal physiology. *An. Rev. Plant Phys. Pl. Mol. Biol.* 41:55–75.

Mellgren, R.L. 1987. Calcium dependent proteins: an enzyme system at cellular membranes. *FASEB Jour.* 1:110–115.

Michalczuk, B., Goszcynska, M., Rudnicki, R. and Halevy, A.H. 1989. Calcium promotes longevity and bud opening in cut rose flowers. *Israel Jour. Bot.* 38:209–15.

Michelangeli, F., Grimes, E.A., East, J.M. and Lee, A.G. 1991. Effects of phospholipids on the function of Ca^{2+}/Mg^{2+} ATPase. *Biochemistry* 30:342–51.

Minorsky, P.V. 1989. Temperature sensing by plants: a review and hypothesis. *Plant Cell Environ.* 8:75–94.

Møller, I., Kay, C.J. and Palmer, J.J. 1986. Chlortetracycline and the transmembrane potential of the innter membrane of plant mitochondria. *Biochem. Jour.* 237:765–71.

Murachi, T., Tanaka, M., Hatanaka, M. and Murakami, T. 1981. Intercellular calcium dependent protease (calpain) and its high molecular weight endogenous inhibitor (calstatin). *Adv. Eng. Reg.* 19:407–24.

Muto, I. 1982. Distribution of calmodulin within wheat leaf cells. *FEBS Lett.* 147:161–4.

Nur, T., Ben Arie, R., Lurie, S. and Altman, A. 1986. Involvement of divalent cations in maintaining cell membrane integrity in stressed apple fruit tissue. *Jour. Pl. Phys.* 125:47–60.

Raven, J.A. 1983. Do plant photoreceptors act at the membrane level? *Philos. Trans. Roy. Soc. London B* 303:403–17.

Reynolds, T.L. 1990. Interaction between calcium and auxin during pollen androgenesis in anther cultures of *Solanum carolinense L. Plant. Sci.* 72:109–14.

Roux, S.J. and Slocum, R.D. 1982. Role of calcium in mediating cellular function in higher plants. pp. 409–53. In: Ed. W.Y. Cheung. *Calcium and Cell Function Vol. 3.* Academic Press, New York.

Schumaker, K.S. and Sze, H. 1984. A Ca^{2+}/H^+ antiport system driven by the proton electrochemical gradient of a tonoplast H^+ ATPase from oat roots. *Pl. Phys.* 74:1111–7.

Schumaker, K.S. and Sze, H. 1990. Solubilization and reconstitution of the oat root vacuolar H^+/Ca^{2+} exchanger. *Pl. Physiol.* 92:340–5.

Schwartz, A., Ilan, N. and Granz, D. 1988. Calcium effects on stomatal movement in *Commelina communis.* L. *Pl. Phys.* 87:583–7.

Serhan, C., Anderson, P., Goodman, E., Dunham, P. and Weissman, G. 1981. Phosphatide and peroxidized fatty acids are calcium ionophores. *Jour. Biol. Chem.* 256:2736–41.

Sridhara, S. and Leshem, Y. 1986. Phospholipid catabolism and senescence of pea foliage membranes; parameters of Ca^{2+}:calmodulin, phospholipase A_2 induced changes. *New Phytol.* 102:5–16.

Stow, J. 1989. The involvement of calcium ions in maintenance of apple fruit tissue structure. *Jour. Exp. Bot.* 40:1052–7.

Tanaka, T., Ohmara, T. and Hiduka, H. 1982. Hydrophobic interaction of the Ca:calmodulin complex with calmodulin antagonists: naphatalenesulfonamide derivatives. *Mol. Pharmac.* 22:403–7.

Triggle, D.J. 1982. Biochemical pharmacology of calcium blockers. pp. 121–34. In:

Eds. S.F. Flaim and R. Zelis. *Calcium Blockers*. Urban and Schwarzenburg, Baltimore and Munich.

Vaughan, M.A. and Mulkey, T.J. 1985. Effects of calcium entry blockers on root elongation and microsomal calcium uptake in maize roots. 12th Int. Cong. Pl. Growth Sub. Heidelberg. Abs. p. 57.

Weiss, B. and Wallace, T.L. 1980. Mechanisms and pharmacological implications of altering calmodulin activity. pp. 330–79. In: Ed. W.Y. Cheung. *Calcium and Cell Function. Vol. I.* Academic Press, New York.

Wong, P.Y.K. and Cheung, W.Y. 1979. Calmodulin stimulates human platelet phospholipase A_2. *Biochm. Biophys. Res. Comm.* 90:473–80.

9. Membrane-associated phospholytic and lipolytic enzymes

The metabolism of plant membrane lipids is important not only in senescence-related processes but also is part and parcel of functional membrane turnover in active cells. Much of the research on phospholytic enzymes has been concerned with *catabolic* aspects and the possibility that the *phospholipase* family of enzymes which are the major membrane-associated lipolytic ones may also function metabolically is overlooked. The direction of the enzymatic reaction – leading either to phospholipid degradation or conversely to resynthesis – is dependent on degree of tissue hydration. Since most plant tissues are highly hydrated, the direction is generally a catabolic one in which case the "retailoring" mechanism is replaced by another group of enzymes – the *acyl-transferases* (the adjective *acyl* pertaining to the fatty acids that comprise the glycerolipid tails).

However situations may arise where degree of hydration is low, causing an *increased* availability or concentration of enzyme substrate. In such cases phospholipase action is essentially expressed as the synthesis of phospholipids from their respective components and not of de-esterification.

From intensive research conducted over several years at the University of Utrecht in Holland and elsewhere it has now emerged that phospholipases as a group are usually located at the interfaces between the membrane bilayer and the internal cytosolic solution as well as at the interfaces between the bilayer and the extra-membranal solution. According to this concept the lipolytic enzymes possess hydrophobic headgroups which have an especial affinity for the water:lipid interfaces and are able to *surf* or *scoot* from one active site to another.

In terms of biophysical interaction with substrate, some of the phospholipases (A_2, B, C, and D) exhibit dependence not only on standard factors such as substrate concentration, pH, temperature, etc. but also upon a factor which in terms of specific activity is unique – *surface tension* of the phospholipid layer upon which they function (Demel et al., 1975). Optimal activity is obtained only within a narrow

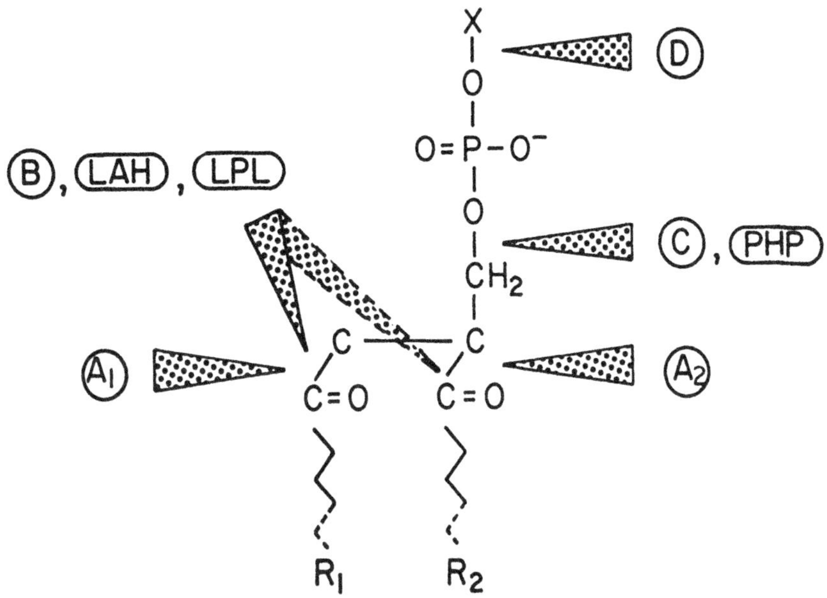

Fig. 9.1. The major groups of phospholipase and lipolytic enzymes. X = choline, ethanolamine, IP, IP_2 or serine. R_1 = usually long chained saturated fatty acids – palmitic or stearic acid (16:0 or 18:0). R_2 = usually long chained unsaturated fatty acids, linoleic or linolenic acid (18:2 or 18:3) in plants, or arachidonic acid (20:4) in mammals. LPL = lysophospholipase; PHP = phosphatidate-phosphatase (* the substrate for this enzyme is phosphatidic acid i.e. the PL lacking the "X" headgroup); LAH = lipolytic acylhydrolase.

range of surface tensions which is 31.0–35.0 mN/m^2 (i.e. about half the surface tension of water) or somewhat lower. This suggests that both excessive bilayer microviscosity on the one hand, and marked rigidity on the other, result in suboptimal enzyme:substrate contact.

Following is a brief description of some of the major membrane-associated lipolytic enzymes. The nomenclature of these enzymes is not always uniform; in the present book the terminology of van den Bosch (1982) is followed (see Fig. 9.1).

9.1. Phospholipase A_1 (PLA_1)

This enzyme de-esterifies the *sn*-1 located fatty acyl chain which often is a saturated one. PLA_1 is widely distributed and apparently, as opposed to most of the other phospholipases, does not require Ca^{2+} for activation.

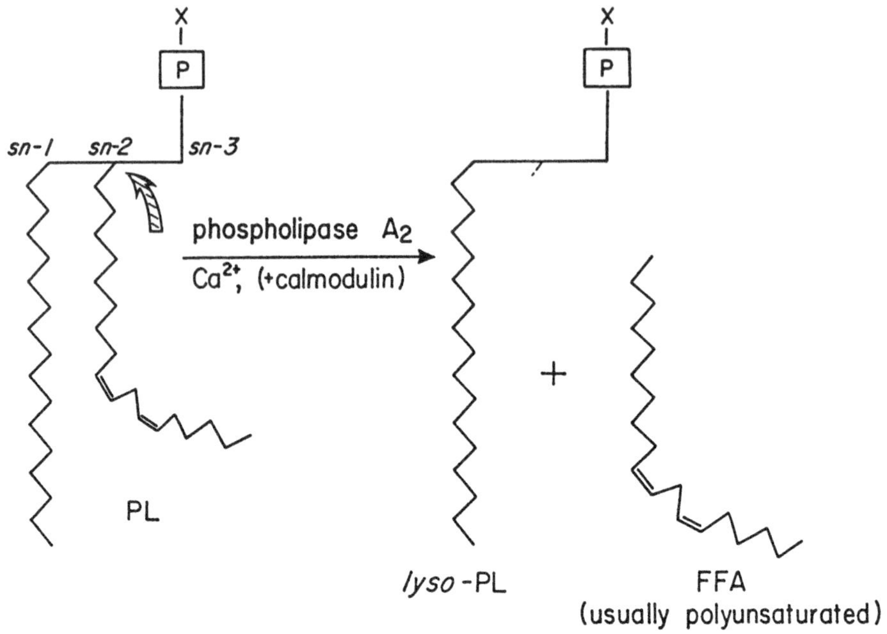

Fig. 9.2. Mode of action of phospholipase A_2.

9.2. Phospholipase A_2 (PLA_2)

PLA_2 de-esterifies the *sn*-2 fatty acyl chain and releases the usually unsaturated fatty acyl chain, the *lyso*-phospholipid remaining (Fig. 9.2). Once bound to the interface, PLA_2 can "scoot" along the bilayer surface and hydrolyze several thousand phospholipid molecules per minute before dissociating from the membrane (Jain et al., 1986). PLA_2's activity is promoted by both Ca^{2+} and calmodulin (Wong and Cheung, 1979). The consensus is that PLA_2 is a group-type and not a single enzyme and that depending on the source, its various isozymes have selective affinity for phospholipid-contained fatty acyl tails of differing chainlengths (Schalwijk et al., 1990). Moreover this enzyme prefers phospholipid aggregates rather than monomers as substrates (Jain et al., 1986).

Occurrence: PLA_2 is of widespread biological occurrence and its activity particularly high in snake, scorpion and bee venoms – part of the toxic effects of bites or stings of these creatures result from lipolytic activity causing localized and topical lysis of blood vessel membranes.

As regarding its occurrence in plants, until recently only indirect evidence suggested presence of PLA_2. However employing Rocket-immunoelectrophoresis and cross reactions with bee-venom PLA_2 antibody immunoglobin, Leshem (1990) and Leshem, Agassi and Av-

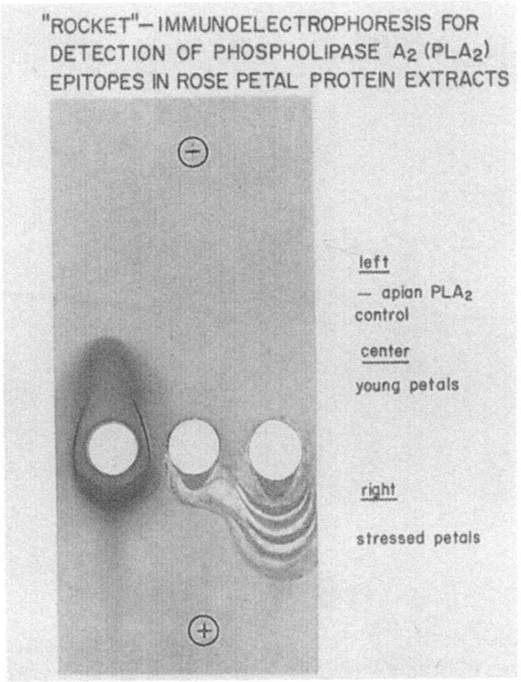

Plate 9.1. Identification of plant phospholipase A_2 in rose petals by 'Rocket' immuno-electrophoresis. Note marked new epitope bands appearing with senescence. Plant protein extracts were challenged with bee venom antibody immunoglobulin (After Leshem et al., 1990). See also Colorplate III pg. 96.

talion (unpublished results) have identified it in rose petals – its appearance, with the formation of additional epitopes, markedly increasing upon heat-stress induced senescence (Plate 9.1 and Colorplate III). Furthermore the PLA_2 antibodies active in rose tissue inhibited linoleic and linolenic (18:2 and 18:3) acid release from soybean phosphatidylcholine substrate as well as from senescing rose petals in which case ethylene production was also inhibited. Moreover specific mammalian PLA_2 inhibitors such as chloropromazine and bromophenolbromide (Borowitz and Montgomery, 1989) elicited similar inhibition of polyunsaturated fatty acid de-esterification in stressed plants (Fig. 9.3).

Mode of action: PLA_2 action on membranes may be likened to an attack with a double-barreled shotgun: not only are two potentially harmful detergents produced – the *lyso*-PL and the released fatty acid (this also being the case for PLA_1) – but also a substrate for subsequent lipoxygenase attack, the latter preferentially acting on a *free* and not PL-bound polyunsaturated fatty acid. The lipoxygenase-mediated

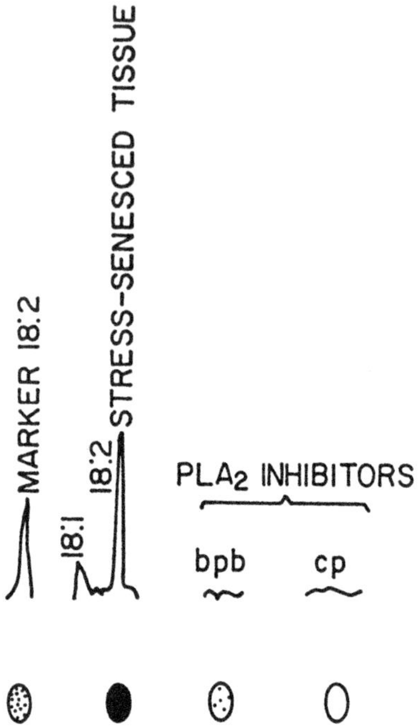

Fig. 9.3. PLA_2 activity of senescing rose petal membranes as affected by PLA_2 inhibitors – bromophenolbromide (bpb) and chloropromazine (cp). *Above* – GLC pattern; *below* – relative I_2 staining of 18:2 RF regions on TLC plates.

peroxidation is an early event in a sequence of reactions which as presently outlined are deleterious or even lethal to cells. It is of more than passing interest that one approach to relief of respiratory stress, allergy and arthritis is the (at present still experimental) pharmaceutical administration of PLA_2 blockers which incipiently prevent formation of lipoxygenase substrates which, amongst other catabolites to be mentioned shortly, ordinarily would lead to stress and malady-associated products such as leucotrienes (Stinson, 1989).

The particular mode of action of the PLA_2-produced *lyso*-phospholipids in terms of membrane biophysics and properties has been discussed in pages 32–4 (*qv*). A further result of PLA_2 action to produce a *lyso*-PL is the formation of a membrane component which, because of stereochemical configuration and natural correspondence of hydrophobic and hydrophilic molecule sections, can afford a special receptor function to senescence evoking abscisic acid (Leshem et al., 1990). See Fig. 9.4.

There is a complex relationship between lipid peroxidation and PLA_2. The process of peroxidation appears to activate PLA_2, and once activated, the enzyme *facilitates* propagation of the peroxidative process. In contrast, if PLA_2 is activated *prior* to *initiation of peroxi-*

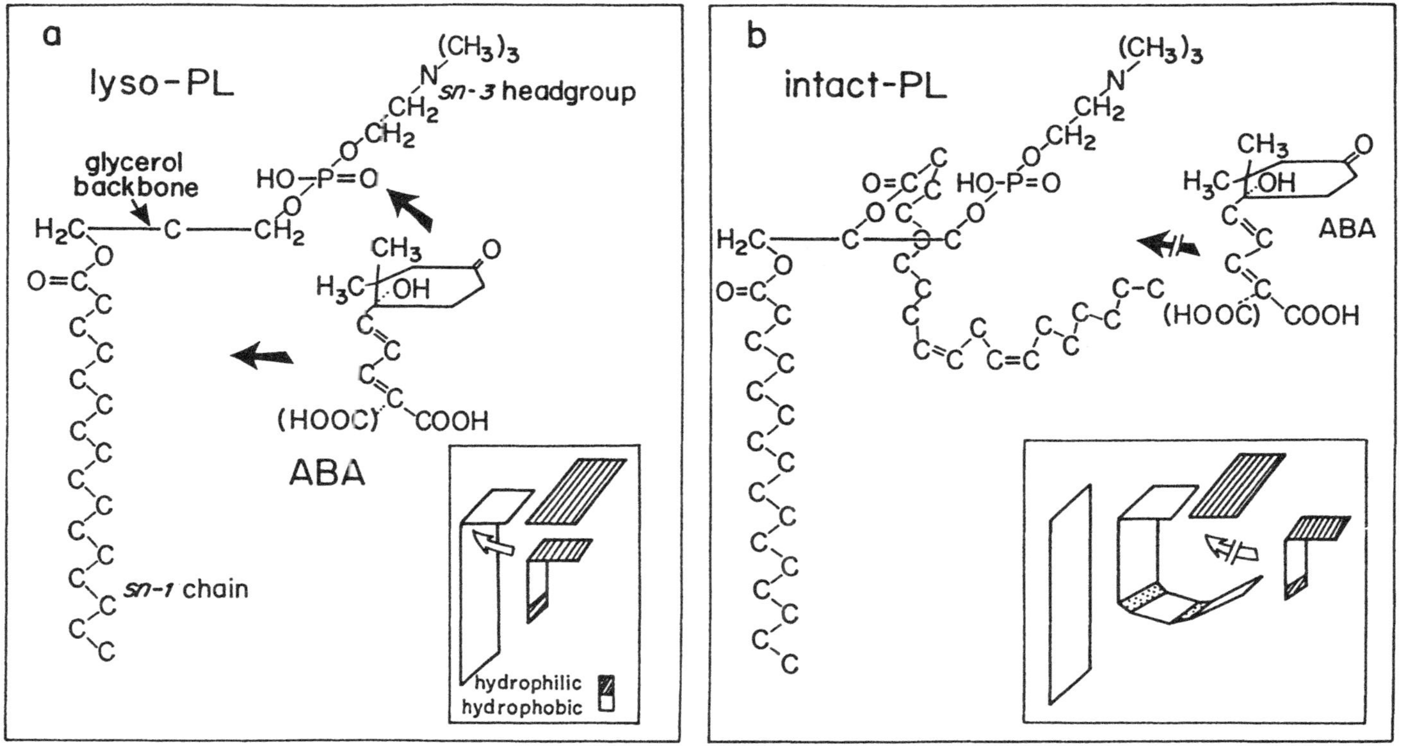

Fig. 9.4. Possible mode of binding of abscisic acid to *lyso*-phospholipids. The latter are produced after stress-induced phospholipase A_2 action. *a:* Mode of binding. *b:* The *intact* PL precludes such binding.

dation, the latter is inhibited. This is because the released fatty acids may migrate into the hydrophilic region of the membrane where they might be less able to interact with other polyunsaturated fatty acids and thus limit propagation of peroxidation (Borowitz and Montgomery, 1989). Moreover PLA_2 activation may be forestalled by facilitating degradation of lipid hydroperoxides by *glutathione peroxidase* (gp) as follows:

$$\text{(glutathione) GSH} + \text{ROOH} \overset{\text{gp}}{\rightarrow} \text{ROH} + H_2O$$

It appears that PLA_2 must be activated before glutathione peroxidase can degrade peroxides since according to the above workers, the enzyme degrades phospholipid-bound hydroperoxides *65,000* times more slowly than it degrades free hydroperoxides.

According to Ansell and Spanner (1982) the two aforementioned enzymes, PLA_1 and PLA_2, actively participate in phospholipid turnover by release of both the *sn*-l and the *sn*-2 fatty acyl sidechains, which concomitantly are rendered available for re-insertion by acyltransferases, this traffic being an integral function of cellular and membrane activity and adaptation to ever-changing environmental conditions.

9.3. Phospholipase B (PLB), lysophospholipase (LPL) and lipolyticacylhydrolase (LAH)

As can be seen in Fig. 9.1 this group of enzymes releases *both* fatty acyl tailgroups. There is a certain degree of uncertainty as to the individual identity of each of these three enzymes, but at present it is commonly held that they are separate entities. As a group they are somewhat heterogenous and certain forms prefer long-tailed fatty acyl chains as substrate while others have greater affinity for short-tailed ones. Their activity has been detected in plants as well as mammals, and they function optimally when both Ca^{2+} and calmodulin are present.

9.4. Phospholipase C (PLC)

As can be seen in Fig. 9.1, and Fig. 12.2 the upshot of PLC's action is the release of the phospholipid polar headgroup, the final products being *diacylglycerol* (DAG) and a *phosphate mono-ester*. As in the case of PLA_2 it appears that PLC is not a single and uniform enzyme but that several types exist, each with a selective preference to specific substrates e.g. PC, PE, PS, PI or PIP_2. Because of its pivotal importance in the process of *signal-transduction* and the release of DAG, this enzyme is commanding widespread interest, and in view of this, it will only be summarily discussed in the present chapter – its physiological role being further detailed in Chapter 12 dealing with the signal transduction process. Here we only mention that its breakdown product, DAG, is one of the most potent biological Ca^{2+} chelators and that the latter are instrumental in protein kinase activation (also discussed in greater detail in Chapter 12). Characterization of PLC in plants has revealed its distribution in a wide variety of species, and within a given species – in various plant organs (Irvine et al., 1980, DiNola and Mayer, 1986). PLC, like PLA_2 requires Ca^{2+} and calmodulin for optimal performance.

9.5. Phospholipase D (PLD)

This enzyme de-esterifies the bond between the *phosphatidic acid* and the *secondary* alcohol sections of PL molecules – these alcohols including choline, ethanolamine, serine and inositol. PLD was initially identified in plants and only later in the other biological kingdoms. This enzyme requires Ca^{2+}, but not calmodulin, for activation. Over and above its general role in membrane turnover we again mention the claim, which as yet remains to be substantiated, that the PLD product, phosphatidic acid, acts as a specific Ca^{2+} ionophore (Putney et al., 1981), the implications of which were discussed in the previous chapter.

9.6. Phosphatidate-phosphatase (PHP)

PHP resembles PLC in action (cf. Fig. 9.1) in that it releases DAG, but differs in that its substrate is PL lacking the secondary alcohol headgroup bonded to the phosphate neck section. Its enzymatic action therefore leads to the formation of DAG and a phosphate group. At present most of the information available on PHP comes from animal studies; however it has been identified in plants where it appears to be associated with membrane fractions (Paliyath and Thompson, 1987; Morré et al., 1990). Concerning its Ca^{2+} requirement – in animal studies, interestingly this ion has been found not only to not promote, but to *inhibit* (Brindley, 1984). To the best of our knowledge this point has not been addressed in plants.

9.7. Lipoxygenase (LOX)

a. Occurrence and distribution: LOX (formerly designated *linoleate oxygen oxido-reductase*) is a group of non-heme containing enzymes catalyzing the stereospecific dioxygenation of polyunsaturated fatty acids (PUFA's) containing a *cis-cis* 1–4 pentadiene system, to produce conjugated *cis-trans*-diene hydroperoxyl derivatives (Fig. 9.5). Like PLD, LOX was initially detected in plants and only later in other biological kingdoms. The PUFA substrates of LOX include linoleic and linolenic acids (18:2 and 18:3) which are present mainly at the *sn*-2 sites of plasmalemmal PLs and at both the *sn*-1 and *sn*-2 sites in chloroplast membranes. As to its subcellular location, LOX has been detected in thylakoid lamellae, somewhat less in chloroplast stroma, as well as in microsomal membranes, the cytosol, vacuole and in lysosomes. In plants, several different isozymes have been identified varying somewhat in temporal order of occurrence, structure, pH optima, and Ca^{2+} requirements, but similar in their basic mode of action (Altschuler et al., 1989; Poca et al., 1990). It is generally held that in order to obtain full expression of LOX potential, release of the FA acyl from the glycerolipid is a prerequisite. However, this contention is at present subject to at least a certain degree of revision and questioning since some of the minor soybean isozymes (2, 3 and 4) **do** possess the ability to act on unliberated PUFA. However, the major LOX isozyme, #1, which comprises ca 70% of total LOX in this species, functions only very weakly on PL-contained PUFA, its activity being increased 1000× when acting upon the free fatty acid (Sofer and Grossman, personal communication).

b. LOX specificity: As stated above LOX specifically acts on the *cis-cis* unsaturated diene (C═C) bonds in PUFA's. The question that arises is – if LOX reacts equally with *all* of fatty acyl-contained C═C bonds, or does it preferentially catalyze only the C═C bonds present at a location within the PUFA? In Chapter 2 it was explained at length that such bonds, as for example in linolenic acid in plants may be at the Δ-9, 12 and 15 while in arachidonic acid typical to mammalian PLs they are located at Δ-5, 8, 11 and 14 sites or as in the long chained PUFA docosapentaenoic acid (DPA) at Δ-4, 7, 10, 13 and 16. In view of the relevance of PL catabolism in the *phosphatidylinositide pathway*

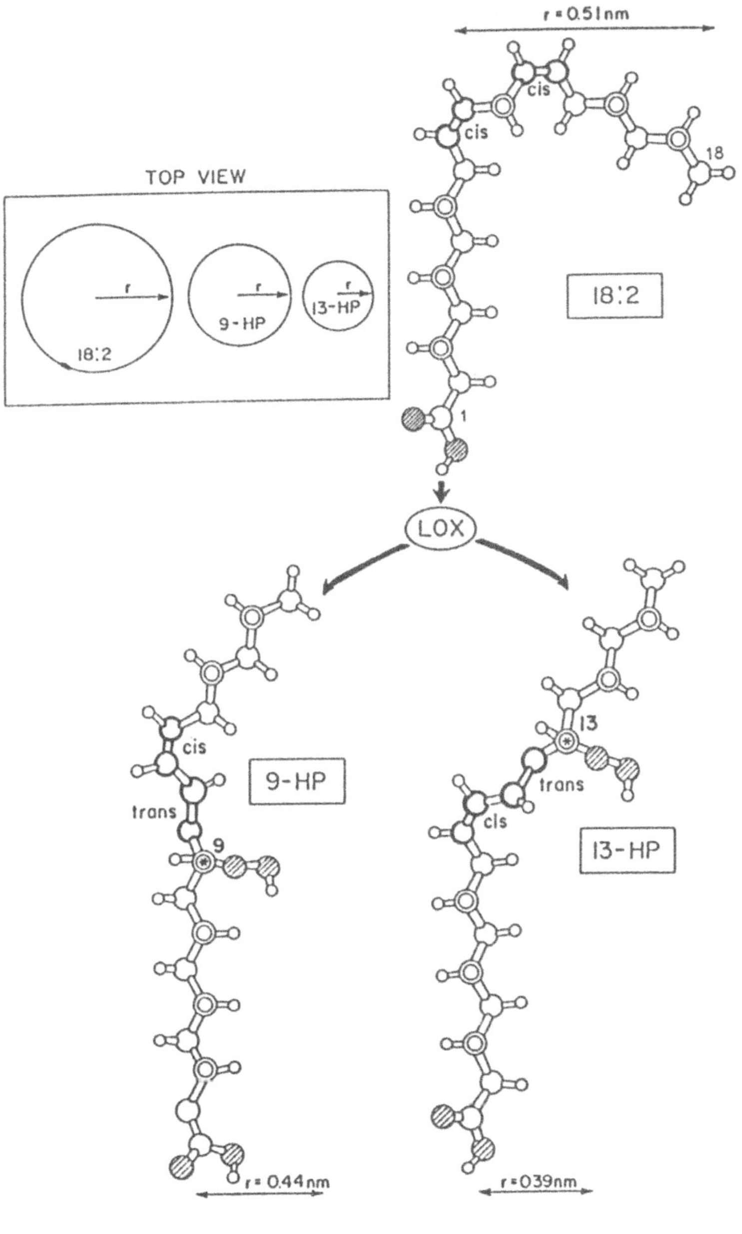

Fig. 9.5. Mode of LOX action. The immediate products are linoleic (or linolenic acid)-9, or -13 hydroperoxides (HP's) in which the *cis-cis* double bond is converted to either a *cis-trans* one as in the 13-HP, or to a *trans-cis* one as in the 9-HP. These changes *reduce* the molecular radius of the 18:2 "backbone". In the case of the 13-HP which is believably the more physiologically active one in the membrane, the molecular area is reduced by half and approximates that of 18:1. This potentially induces greater membrane rigidity by allowing packing of more molecules into a given area.

and in the *phosphatidyl-linoleic cascade* which will be described in Chapter 12 (see also Fig. 12.2) and of LOX's participation in formation of leucotrienes which are associated with tumor and cancer development, as well with certain aspects of the processed food industry, this question has been intensely studied. The pattern that has emerged is that plant LOX differs from animal LOX's and that it is specific to the C=C *cis-cis* bonds specific to plant PUFA's and is ineffective or acts only weakly in catalyzing the other bonds occurring elsewhere in animal specific PUFA's, and vice versa.

This section does not aim to provide a complete description of LOX metabolism in biology and is primarily geared to plant LOX, but at this point it is pointed out that several different animalian forms of LOX have been found, differing from each other in specific substrate, site and location of action.

c. Mode of action: According to Galliard (1980) the LOX reaction involves formation of free radicals and is dependent upon interaction between the enzyme, its substrate and O_2, iron ions also being necessary. The immediate products of LOX action are the *hydroperoxides* of the respective fatty acids, and these hydroperoxides themselves may subsequently serve as a substrate for further LOX catalysis to obtain additional breakdown adducts (cf. Fig. 9.5). Lipoxygenase preparations from potato tubers convert linolenic acid not only into 9-hydroperoxide, but also to several more polar metabolites. These include hydroxy/hydroperoxy, dihydroperoxy, 9-oxo-16-hydroperoxy- and 9-hydroperoxy-16-oxo- derivatives of linolenate with three conjugated double bonds which may account for >70% of the total (Tarchevsky et al., 1990; Grechkin et al., 1991).

The two most well studied lipoxygenases are the soybean seed and potato tuber types. A common feature of these and other lipoxygenases of *seeds* is that the divalent Fe^{2+} appears to be part of the enzyme – believably associated with histidine-rich areas in the monomeric protein, whereas *foliage*-contained lipoxygenases contain the trivalent Fe^{3+} ion. As shown in Fig. 9.5 the immediate products of plant LOX are the linoleic, (or to a lesser extent the linolenic) *hydroperoxides*, the site of peroxidation between either the 9- or the 13-C atom of the 18:2 backbone.

In terms of category of hydroperoxide release, LOX from various species, or within the same botanical species but from different plant organs, have been studied in various laboratories and the overall presently-known scheme is presented in Table 9.1. From this table it is seen that soybean LOX, especially the major isozyme 1, primarily

Table 9.1. Percentage distribution of 18:2 hydroperoxides (HP's) produced by plant lipoxygenases from various sources

LOX source	9-HP	13-HP	Remarks
Seeds			
Soybean isozyme 1	–	>95	Contains Fe^{2+}
Soybean isozyme 2	50	50	Contains Fe^{2+}
Soybean isozyme 3	50	50	Contains Fe^{2+}
Soybean isozyme 4	50	50	Contains Fe^{2+}
Flax	30	70	
Corn	90	10	
Barley	>95	–	
Tubers			
Potato	>95	–	Also some 9, 16 double HP
Foliage			
Soybean	–	80–90	Contains Fe^{3+}
Tea	–	>80	Contains Fe^{3+}

(Data kindly provided by Professor Gerrit Veldink, Dept. of Bioorganic Chemistry, University of Utrecht, The Netherlands).

produces 13-HP whereas potato and barley LOX produce 9-HP. In terms of membrane turnover and physiology, the hydroperoxidation at the 13 site is more meaningful since being located further down along the fatty acyl chains of membrane it is more prone to potential LOX attack than the 9 site. As graphically depicted in Fig. 9.5 it can be seen that hydroperoxidation may induce membrane rigidity by markedly decreasing molecular radius.

During the catabolic process an array of oxy-free radicals appear: at first the PUFA chain may become "free-radicalized" to form the *PUFA free radical* and subsequently the *peroxyl* ($ROO^{\cdot}$), *alkoxyl* ($RO^{\cdot}$) and the *superoxide* ($O_2^{\cdot}$) free radicals. Due to the formation of these free radicals, LOX action is potentially harmful to membranes since leaks may form and/or rigidity increase. LOX induces rigidity since more LOX-treated PLs may be compressed into a given membrane area than in membranes not subjected to LOX catalysis (Bamberger et al., 1989). Senescence of various plant organs has been accompanied by marked increase of LOX levels (Grossman and Leshem, 1978; Bousquet and Thimann, 1984; Lynch et al., 1985) however the enhanced LOX activity may also be encountered in non-senescence related processes where fatty acid storage products are being released for utilization as for example in germinating seeds (Peterman and Siedow, 1985). The cellular compartment which in senescing pea foliage manifested LOX activity was the microsomal one, which is mainly plasmalemma (Sridhara and Leshem, 1986).

d. LOX breakdown products: The said hydroperoxides undergo further catabolism and give lead to the production of aldehydes such as malondealdehyde, volatile hydrocarbons – ethane, pentane and ethylene – and also to the appearance of *jasmonic acid* or its methyl derivative. In wounded or damaged tissues there is a very rapid and massive increase of C-6 aldehyde production, the different LOX isozymes manifesting different aldehyde-producing capacities, #2 isozyme usually being the most active in this respect (Hildebrand et al., 1990). The above LOX catabolites are important in imparting culinary quality to processed vegetable produce and also in bleaching of flour. But, as in the case of processing tomato paste for ketchup, they may induce bleaching of the red coloration and lead to the formation of "off-taste" compounds (Davood and Biacs, 1988). Similar discoloration may occur during processing of red pepper powder for paprika.

Research on both mammalian and plant tissues indicates that peroxidation of membrane lipids may cause accumulation of fluorescent lipid compounds resembling the fluorescent pigment *lipofuscin* which builds up in the myocardium surrounding the heart and which serves as a warning of incipient cardial malfunction. Detailed reviews on this subject have been written by Merzlyak et al. (1983) of the Moscow University and by Duggelin et al. (1988), and the interested reader can there pursue the topic further. As for malondealdehyde which is intricately linked to the above described sequence of catabolic events, the reader's attention is called to the section dealing with "cross-linking" of proteins which appears in Chapter 5.

The typical odors of freshly cut cucumbers, freshly mowed lawns and of pressed tea-leaves also stem from LOX-produced catabolites these including aldehydes such as *hexanal*, *nonenal*, and the alcohol *hexenol* (Hatanka et al., 1987). LOX also apparently participates in degradation apo-carotenoids formed as by-products of cleavage (suggested to occur at the chloroplast membrane) of the immediate precleavage precursor of ABA – 9′-*cis*-neoxanthin (Parry and Horgan, 1991). Fig. 9.6 illustrates how the initial LOX breakdown products may further be catabolized.

Three main pathways are followed:

i) Ketone formation mediated by *hydroperoxide-isomerase*.
ii) *Aldehyde* formation by *hydroperoxide-lyase*. The latter enzyme is a 55 kD protein whose relative activities for hydroperoxides of linoleic, α-linolenic and γ-linolenic acids were determined as 100, 843 and 35% respectively. Hydroperoxide cleaving activity is highly sensitive to antioxidants such as *butylated hydroxyanisole* (BHA) (Hatanka et al., 1990). Recent developments and im-

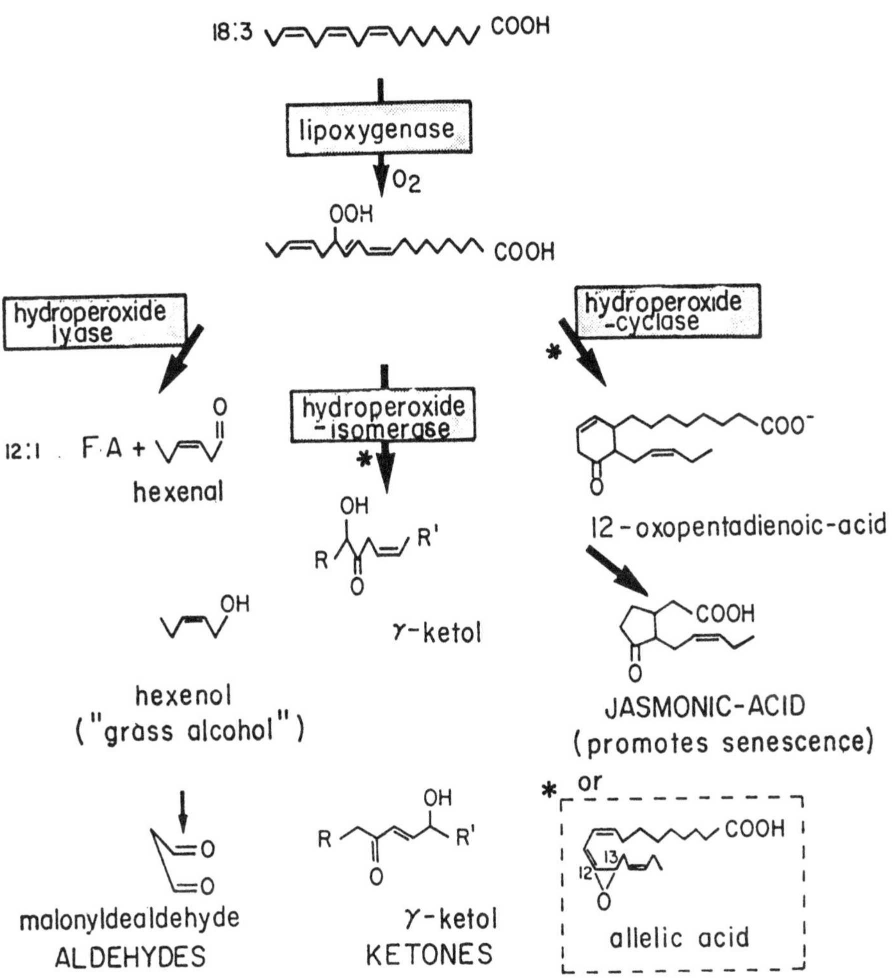

Fig. 9.6. Lipoxygenase breakdown products and pathways (After Vick and Zimmerman, 1987). The asterisk * indicates an alternative route, not via the cyclase or isomerase, according to Brash et al. (1988).

proved spectrophotometric assays of hydroperoxide-lyase now allows hydroperoxide-lyase determination with a high degree of accuracy (Vick, 1991).

iii) "Hairpin bending" of the initial LOX catabolite by *hydroperoxide-cyclase* to produce *12-oxophytodienoic acid* and subsequently *jasmonic acid* derivatives which as mentioned are associated with senescence and possibly also with phytopathological infestation. The plant cyclase is very much akin to the mammalian *cyclo-oxygenase* to which a key regulatory function is ascribed in the "arachadonic acid pathway" and in the endogenous production, in animals, of the *prostaglandin* hormones (cf. also Fig. 12.2 in Chapter

12) which in their "hairpin-bent" structure resemble jasmonic acid (Sembdner et al., 1990; Ueda et al., 1991).

The above mentioned scheme as graphically depicted in Fig. 9.6 is essentially based on the information aptly coordinated by Vick and Zimmerman (1987). However Brash et al. (1988) have recently indicated that steps "ii" and "iii" above do not necessarily involve, respectively, hydroperoxide isomerase and cyclase, in order to obtain the end products which as seen in Fig. 9.6 are aldehydes and jasmonic acid. In research conducted on flaxseed tissue which is rich in linoleic and linolenic acids, Brash and colleagues reported that the initial LOX catabolite – the fatty acid hydroperoxide – may convert to the extremely volatile *epoxyl* adduct *allene oxide* which may subsquently convert to the aforementioned end products.

The quantitative ratio relationship between enzymatic conversions and the allene oxide pathway is at present not known. We conclude this section with the surmise that if indeed the cyclase is not necessary for "bending" the straight chained initial LOX catabolite, then this finding has pertinence not only to plant function but also possibly to the formation of leucotrienes, prostaglandins, thromboxanes and similar products in mammals. A further discussion of LOX, including aspects beyond the topics dealt with in this book, is given by Vick and Zimmerman (1987).

9.8. Connection between protease and phospholipase activities

In trials conducted on potato tubers it has been shown that if tissues are subjected to proteolysis prior to phospholytic activity, the latter is considerably enhanced (Moreau and Morgan, 1988). This research assayed overall phospholytic action without distinguishing between the individual phospholipases, but is nevertheless a most salient indication that the factor of proteolysis should be taken into consideration as a preliminary step to optimal digestive action of phospholipases. Moreover this finding possibly confirms the contention of most dietary practitioners that the human digestive tract is equipped with the ability to cope with a *mixture* of basic food categories – this contrasting with the opinion that for optimal digestion it is advantageous to partake of only one class of nutrient at a time.

9.9. References and recommended reading

Altschuler, M., Grayburn, W.S., Collins, G.B. and Hildebrand, D.F. 1989. Developmental expression of lipoxygenase in soybeans. *Pl. Sc.* 63:151–8.

Ansell, G.B. and Spanner, S. 1982, Phosphatidylserine, phosphatidylethanolamine and phosphatidylcholine. pp. 1–50. In: Eds. J.N. Hawthorne and G.B. Ansell. *Phospholipids*. Elsevier, Amsterdam.

Bamberger, E.S., Alter, M., Landau, E.M. and Leshem, Y.Y. 1989. Biophysical effects of superoxide on surface parameters of a model membrane. pp. 113–6. In: Eds. O. Hayaishi, E. Niki, M. Kondo and T. Yoshikawa. *Medical, Biochemical and Chemical Aspects of Free Radicals*. Elsevier, Amsterdam.

Borowitz, S.M. and Montgomery, C. 1989. The role of phospholipase A_2 in microsomal lipid peroxidation induced with *t*-butyl hydroperoxide. *Biochem. Biophys. Res. Comm.* 158:1021–8.

Bousquet, J.F. and Thimann, K.V. 1984. Lipid peroxidation forms ethylene from ACC and may operate in leaf senescence. *Proc. Nat. Acad. Sci. USA* 81:1724–7.

Brash, A.R., Baertschi, S.W., Ingram, C.D. and Harris, T.M. 1988. Isolation and characterization of natural allene oxides: unstable intermediates in the metabolism of lipid hydroperoxides. *Proc. Nat. Acad. Sci. USA* 85:3382–6.

Brindley, D.N. 1984. Intracellular translocation of phosphatidate phosphohydralyse and its possible role in the control of glycolipid synthesis. *Prog. Lip. Res.* 23:115–33.

Davood, H. and Biacs, P.A. 1988. Some properties of tomato lipoxygenase. *Act. Aliment.* 17:53–65.

Demel, R., Geurt, W., van Kessel, A., Zwaal, R., Roelofsen, B. and van Deenen, L.M. 1975. Relations between various phopholipase actions on human red cell membranes and the interfacial phospholipid pressure in monolayers. *Biochim. Biophys. Acta* 406:97–107.

DiNola, L. and Mayer, A.M. 1986. Effect of temperature on glycerol metabolism in membranes and on phospholipase C and D of germinating pea embryos. *Phytochem.* 25:2255–9.

Duggelin, T., Bartlik, K., Gut, H., Matile, P. and Thomas, H. 1988. Leaf senescence in *Festuca pratensis*: accumulation of lipofuscin-like compounds. *Physiol. Plant.* 74:131–6.

Galliard, T. 1980. Degradation of acyl lipids: hydrolytic and oxidative enzymes. pp. 85–116. In: Ed. P.K. Stumpf. *The Biochemistry of Plants. Vol. 4.* Academic Press, New York.

Grechkin, A.N., Kuramshin, R.A., Safonava, E.Y., Yefremov, Y.J., Latypov, S.K., Ilyasov, A.V. and Tarchevsky, I.A. 1991. Double hydroperoxidation of α-linolenic acid by potato tuber lipoxygenase. *Biochim. Biophys. Acta*, 1081:79–84.

Grossman, S. and Leshem, Y. 1978. Lowering of endogenous lipoxygenase activity in *Pisum sativum* foliage by cytokinin as related to senescence. *Physiol. Plant* 43:359–62.

Hatanka, A., Kajiwara, T. and Sekiya, J. 1987. Biosynthetic pathway for C-6 aldehydes formation from linolenic acid in green leaves. *Chem. Phys. Lipid* 44:341–61.

Hatanka, A., Kajiwara, T. and Matsui, K. 1990. Enzymatic oxygenative-cleavage reaction of 2-linolenic acid in tea leaves. pp. 295–7. In: Eds. P.J. Quinn and J.L. Harwood. *Plant Lipid Biochemistry. Structure and Utilization.* Portland Press, Colchester and London.

Hildebrand, D.F., Zuang, H., Hamilton-Kemp, I.R. and Loughrin, J.H. 1990. pp. 307–9. In: Eds. P.J. Quinn and J.L. Harwood. *Plant Lipid Biochemistry. Structure and Utilization.* Portland Press, Colchester and London.

Irvine, R.F., Letcher, A.J. and Dawson, R.M.C. 1980. Phosphatidyl-phosphodiesterase in higher plants. *Biochem. Jour.* 192:279–83.

Jain, M.K., Rogert, R.J., Maracek, J.F., Ramirez, F. and Eibel, H. 1986. Effect of the structure of phospholipid on the kinetics of intravesicle scooting of phospholipase A_2. *Biochim. Biophys. Acta* 860:461–74.

Leshem, Y. 1988. Plant senescence processes and free radicals. *Free Rad. Biol. Med.* 5:39–49.

Leshem, Y.Y. 1990. Evidence for the presence and mode of action of a membrane-associated plant phospholipase A_2. pp. 53–5. In: Eds. P.J. Quinn and J.L. Harwood. *Plant Lipid Biochemistry. Structure and Utilization*. Portand Press, Colchester and London.

Leshem, Y.Y., Cojocaru, M., Mergel, S., El-Ani, D. and Landau, E.M. 1990. A biophysical study of abscisic acid interaction with a phospholipid membrane component. *New Phytol.* 116:487–98.

Leshem, Y.Y., Halevy, A.H. and Frenkel, C. 1986. *Processes and Control of Plant Senescence*. Elsevier, Amsterdam. 215 pp.

Lynch, D.V., Sridhara, S. and Thompson, J.E. 1985. Lipoxygenase generated hydroperoxides account for the non-physiological features of ethylene formation from ACC by microsomal membranes of carnations. *Planta* 164:121–5.

Merzlyak, M.A., Rumyantseva, V., Shevyrova, V. and Gusev, M. 1983. Further investigation of liposoluble fluorescent compounds in senescing plant cells. *J. Exp. Bot.* 34:604–9.

Moreau, R. and Morgan, P. 1988. Proteolytic activation of a lipolytic enzyme activity in potatoes. *Plant Sci.* 55:205–11.

Morré, D.J., Marmé, D.M. and Penel, C. 1990. Calmodulin-calcium stimulation of a 2,4-D responsive phospholipase of soybean membranes. *Protoplasma* (In press).

Paliyath, G. and Thompson, J.E. 1987. Calcium and calmodulin regulated breakdown of phospholipid by microsomal membranes from bean cotyledons. *Pl. Physiol.* 83:63–8.

Parry, A.D. and Horgan, R. 1991. Carotenoids and abscisic acid (ABA) biosynthesis in higher plant. *Physiol. Plant.* 82:320–6.

Peterman, T.K. and Siedow, J.N. 1985. Behavior of lipoxygenase during establishment, senescence and rejuvenation of soybean cotyledons. *Plant. Physiol.* 78:690–5.

Poca, E., Rabinovitch-Chable, H., Cook-Moreau, J., Pages, M. and Rigaud, M. 1990. Lipoxygenase from *Zea mays*. Purification and chemical characteristics. *Biochim. Biophys. Acta* 1045:107–14.

Putney, J.W., Poggioli, P. and Weiss, S.G. 1981. Receptor regulation of calcium release and calcium permeability in parotid glands. *Phil Trans. Roy. Soc. Lond. Ser. 13.* 296:36–45.

Schalkwijk, C.G., Mark, F., van den Bosch, H. 1990. Studies on the acyl chain selectivity of cellular phospholipase A_2. *Biochim. Biophys. Acta* 1044:139–40.

Sembdner, G., Meyer, A., Miersch, O. and Brückner, C. 1990. Metabolism of jasmonic acid. pp. 374–9. In: Eds. R.P. Pharis and S.B. Rood. *Plant Growth Substances 1988.* Springer Verlag, Berlin, Heidelberg.

Sridhara, T. and Leshem, Y. 1986. Phospholipid catabolism and senescence of pea foliage membranes; parameters of Ca^{2+} calmodulin phospholipase A_2 induced changes. *New Phytol.* 102:5–16.

Stinson, S.C. 1989. Better understanding of arthritis leading to new drugs to treat it. *Chem. Eng. News* 67:37–70.

Tarchevsky, I.A., Kuramshin, R.A. and Grechkin, A.N. 1990. Conversion of α-linolenate into conjugated trienes and oxotrienes by potato tuber lipoxygenase. pp. 298–300. In: Eds. P.J. Quinn and J.L. Harwood. *Plant Lipid Biochemistry, Structure and Function.* Portland Press, London and Colchester.

Ueda, J., Mizumoto, T. and Kato, J. 1991. Quantitative changes of abscisic acid and methyl jasmonate correlated with vernal leaf abscission of *Ficus superba* var. *japonica*. *Biochem. Physiol. Pflanz.* 187:203–10.

van den Bosch, H. 1982. Phospholipases. pp. 313–58. In: Eds. J.N. Hawthorne and G.B. Ansell. *Phospholipids*. Elsevier, Amsterdam.

Vick, B.A. 1991. A spectrophotometric assay for hydroperoxide-lyase. *Lipids*. 26:315–20.

Vick, B.A. and Zimmerman, D.C. 1987. Oxidative systems for modification of fatty acids: the lipoxygenase pathway. pp. 54–90. In: Eds. P.K. Stumpff and E.E. Conn. *The Biochemistry of Plants. Vol. 9.* Academic Press, London, New York.

Wong, P.Y.K. and Cheung, W.Y. 1979. Calmodulin stimulates human platelet phospholipase A_2. *Biochem. Biophys. Res. Comm.* 90:473–80.

10. Response of plant membranes to chilling and freezing

R.L. SHEWFELT

Chilling and freezing injuries are the major consequences of low temperature stress in plants. Chilling injury is incurred in susceptible species at temperatures above the freezing point but below a critical threshold temperature. Freezing injury is incurred as ice crystals form within the tissue although some species are more likely to recover from freezing and thawing than others. Acclimation is a developmental process in response to environmental conditions which leads to increased tolerance to low temperature stress. Chilling and freezing interfere with both photosynthetic and non-photosynthetic processes in the cell. Chilling injury leads to reduced crop yields in the field and compromised quality of harvested products during storage and handling. Freezing injury can result in plant death and unmarketable product.

The reasons why some species are more susceptible to low temperature injury are not clear, but most scientific evidence points to the importance of cellular membranes. This chapter focuses on the potential role of biophysical changes of membranes in the etiology of low-temperature injury as these changes are affected by modification of lipid composition and as they potentiate activity of membrane-bound proteins.

10.1. Proposed mechanisms of plant cell response to low temperatures

The early concept that freezing injury is strictly the result of physical disruption of tissues by ice crystal formation could not explain the wide variation observed in susceptibility of different species to low temperatures. One of the first mechanisms proposed to explain cellular response to low temperatures on a chemical basis suggests that disulfide bonds form between two glutathione molecules or between sulfhydryl groups on adjacent proteins during freezing (Levitt, 1962). According to the theory these bonds result from localized dehydration in the cell bringing proteins together. Upon thawing the bonds remain intact restricting protein structure and thus affecting protein activity. Lowered protein activity in turn leads to metabolic imbalances and general cellular disorder. A more refined version of the hypothesis (Levitt, 1972) incorporates the growing understanding of membrane structure suggesting that disulfide bonds between membrane-bound proteins are responsible for freezing injury as well as for electrolyte leakage. In a subsequent edition of his book Levitt (1980) discounts the sulfhydryl hypothesis as an artifact but suggests that glutathione-glutathione binding may increase susceptibility to oxidase activity. Levitt's important contributions are (1) to distinguish primary reactions associated with tissue response and the resultant secondary effects and (2) to identify proteins as the most likely target of damage with sulfhydryl groups as the most susceptible functional group.

The most influential article on chilling injury research is the publication of the phase-transition hypothesis by Lyons (1973) based on his work with Raison. By this theory, a transition is induced in membrane lipids at the critical threshold temperature from liquid-crystalline to solid gel phases. A membrane in the solid gel phase would be expected to show increased permeability and lead to an increase in activation energy of membrane-bound enzymes. Resultant changes in enzyme activity would lead to metabolic imbalances and accumulation of toxic substances resulting in cell injury and death. Differences in species susceptibility are attributed to differences in the degree of saturation of fatty acids of membrane phospholipids. Mitochondrial membranes, which contain critical respiratory enzymes, are considered the most likely target. A sharp change in the rate of a mitochondrial enzyme as detected by a split Arrhenius plot is considered evidence to support the hypothesis.

The phase-transition hypothesis stimulated much research and criticism with the divergent viewpoints published in a book edited by Lyons et al. (1979). The experimental data suggest that:

(1) phase transitions are not as sharp in mixed phospholipid membranes as the theory suggested,
(2) these transitions are not likely to occur above 0°C,
(3) mitochondrial membranes are not very sensitive to chilling temperatures,
(4) Arrhenius plots provide misleading information on membrane-bound enzymes, and

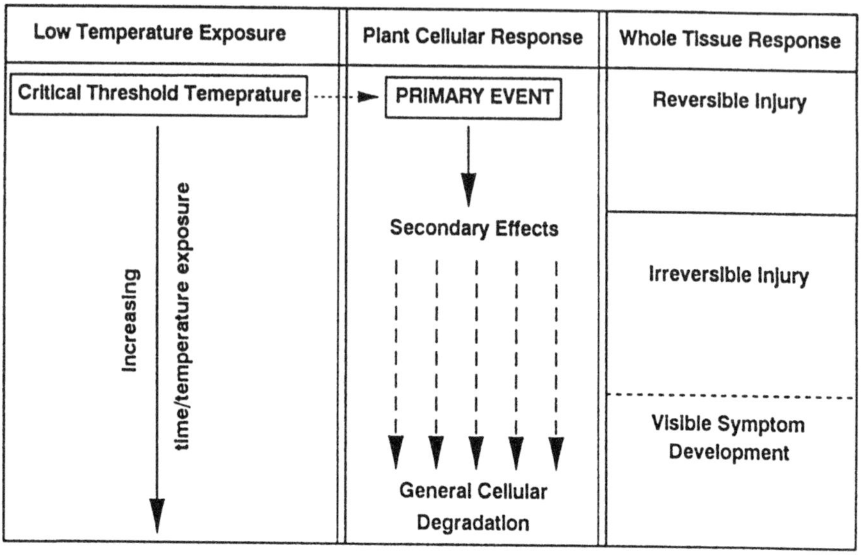

Fig. 10.1. Generalized model for development of low temperature injury in the cell and tissue.

(5) lipid compositional studies do not correlate well with chilling susceptibility.

Despite its deficiencies, Steponkus (1981) concludes in one of the most comprehensive critiques of the phase-transition hypothesis that "the most fundamental tenet that thermal influences on the physical properties of cellular membranes can significantly alter biological activities remains the most attractive explanation for chilling injury." The theory has been successfully modified in the intervening years to remain the primary influence on chilling injury research. The Lyons hypothesis has been extended to provide explanations for senescence and other membrane-associated disorders including freezing.

The use of modifications to explain a broad range of data led Raison and Lyons (1986) to issue a plea for uniform terminology. In this article they clearly partition chilling effects into primary and secondary events. The primary event must occur as the critical threshold temperature is reached, be readily detectable and be reversible during short-term exposure. Secondary effects are derived from the primary event and can lead to irreversibility and symptom development. An illustration of this concept appears in Fig. 10.1. A biophysical change in membranes remains the major candidate for the primary event, but alteration of the conformation of key proteins, the cytoskeleton or calcium balance have also been suggested. To date no primary event has been clearly established in the chilling response. Although Raison and Lyons emphasize that the primary event may differ in different tissue, implicit in their argument is that the key to understanding chilling injury and its control is elucidation of the primary event. Such reasoning suggests an active process that inhibits synthesis or activates degradation is involved.

In whole plants or detached leaves which show evidence of light-induced chilling injury within hours, the relevance of a potential primary event is clear. In harvested tissue where symptom development is not evident for days or weeks, it is conceivable that passive changes in the balance between degradation and synthesis or repair could result in injury. In the latter case, a primary event may not occur and the critical, controllable event could be what is normally considered a secondary response.

From the above discussion it is apparent that there are still questions as to how membranes respond to low temperatures. Despite disagreements, most plant membrane physiologists who study response to low temperatures would agree that biophysical changes in membranes in response to chilling, freezing and thawing are affected by the composition of lipids in the membrane and that these biophysical properties can potentiate activity of membrane-bound proteins. Such a model is presented in Fig. 10.2. The survival of the Lyons hypothesis is due, in part, to the ability to redefine terms while retaining the basic concept. Much of the research on chilling, freezing and acclimation has focused on composition of membrane lipids. This chapter presents the topic of low temperature injury from a biophysical perspective.

10.2. Chilling injury

Experimental evidence suggests that chilling injury can be divided into two separate syndromes that proceed via distinct membrane-mediated mechanisms. One type of chilling injury develops rapidly, either requires or is enhanced by light, affects photosynthetic processes and is readily reversible upon rewarming. The other type proceeds more slowly, affects non-photosynthetic processes and does not usually become evident until after rewarming. Thylakoids are the most likely target of light-enhanced chilling injury while the plasma membrane and tonoplast are suggested targets for dark injury. Recent compositional studies focus on specific molecular species of membrane lipids, lipid hydroperoxides, sterol/phospholipid ratio and phospholipid head groups. Minorsky (1985, 1989) has presented a comprehensive hypothesis whereby both chilling and cold shock-induced membrane injury is triggered by Ca^{2+} passage from the cell's exterior into the cytosol.

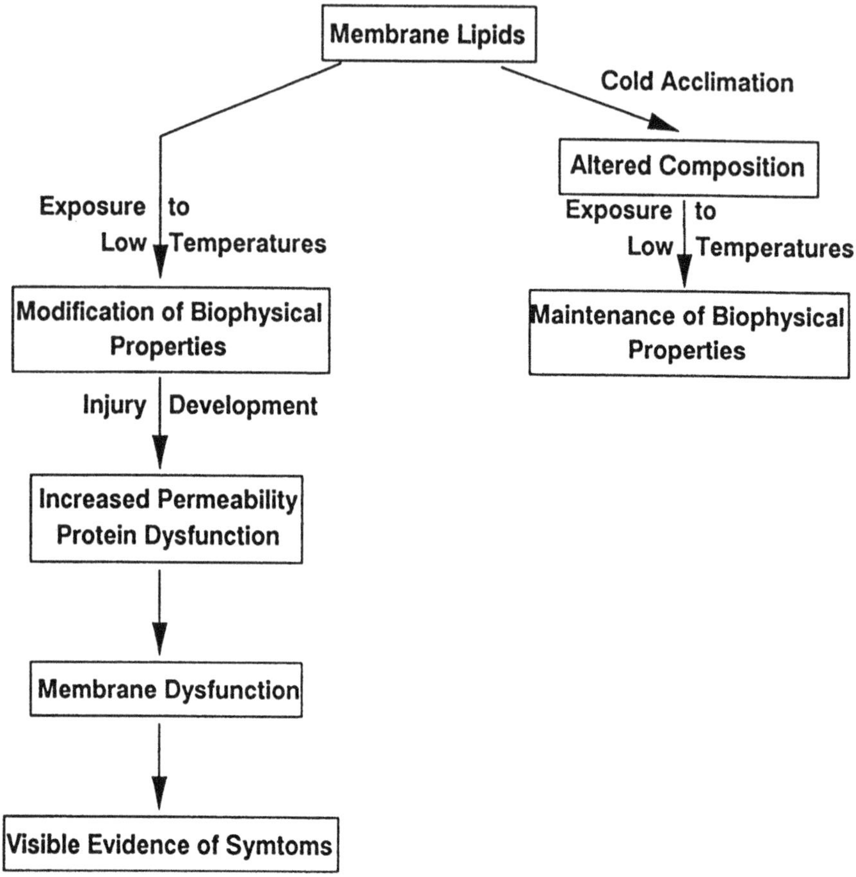

Fig. 10.2. Generalized model to explain the role of membrane lipids in the development of low temperature injury.

The sequence of events envisaged in this hypothesis resembles that outlined in Chapter 12 when describing the "*Ca^{2+}-induced phosphatidyl-linoleyl cascade*". Biophysical studies primarily relate compositional changes to decreased membrane "fluidity." Little characterization of chilling-induced, membrane-mediated changes in protein activity has been conducted.

10.2.1. *Intercellular localization*

Plastids are the primary focus of both in vivo and in vitro studies of light-enhanced chilling injury in plants. Photosynthesis is greatly restricted in whole plants and detached leaves of chilling susceptible species in the presence of light as evidenced by reduction of chlorophyll fluorescence. Photophosphorylation is inhibited by a disruption of photosystem II located in the thylakoids. The effect is readily reversible after short-term chilling exposures upon rewarming. An apparent contradiction between in vivo and in vitro studies has recently been resolved by Wise et al. (1990). Uncoupled photophosphorylation is noted in intact cucumber leaves exposed to chilling in the light and chloroplasts isolated from the stressed leaves. Photophosphorylation is restored in these leaves after 30 min rewarming and chloroplasts isolated from the rewarmed leaves, but it is not restored in chloroplasts isolated from chilled leaves that were not rewarmed. The authors attribute these results to a coupling factor, as yet unidentified, that binds during rewarming in intact tissue but not in isolated chloroplasts.

Ion leakage, a frequently-used measure of chilling injury, implicates the role of the plasma membrane as a primary site of dark injury. Loss of plasma membrane integrity poses catastrophic consequences for the life of a plant cell and the integrity of the adjacent cell wall. Other suggested sites of injury include the tonoplast and mitochondrial membranes.

Both ultrastructural studies and observations of symptoms suggest a sequential development of injury as the duration of exposure continues. Unfortunately in detached tomato fruit the data derived from the two approaches do not support each other. In an ultrastructural study of mature-green fruit held at 5°C, Marangoni et al. (1989) observe a progression of degradation of starch granules in plastids within 2 days; followed by cell vesiculation, dilation of the endoplasmic reticulum, loss of ribosomes and swollen plastids by 7 days; and complete disorganization of the cytoplasm including loss of mitochondria and other organellar features in the cytoplasm after 14 days. During the observation period the plasma membrane remains intact. These results are in stark contrast to symptom development observed in breaker fruit at 4°C. Cheng and Shewfelt (1988) note increased susceptibility of decay and ion leakage indicative of plasma membrane dysfunction after 15 days, followed by an inhibition of ethylene production related to tonoplast dysfunction after 27 days, inhibition of color development related to plastid membranes in 34 days, and maintenance of mitochondrial respiration for more than 35 days. One critical difference in the two studies is that ultrastructural changes are noted at chilling temperatures only while symptoms are observed upon rewarming after chilling, but no explanation has been offered to explain obvious contradictions.

10.2.3. *Composition*

When numerous studies failed to establish a definitive link between fatty-acid composition and chilling sensitivity, the direction of research turned to the role of molecular species of specific lipid classes as reviewed by Thompson (1989) and Lynch (1990). It was suggested that wholesale changes in bulk lipids of membranes are not necessary to change physical properties of the membrane. Clustering of distinctive molecular species might be sufficient to induce localized phase changes within the membrane to affect lipid-protein and lipid-lipid interactions.

Murata et al. (1982) suggest that the distribution of high-melting fatty acids at the *sn*-1 and *n*-2 positions of phosphatidylglycerol (PG) are sufficient to induce phase separations within the membrane. PG is the major phospholipid in thylakoids in which the predominant lipid fraction are the galactolipids. PG is unique among glycerolipids found in membranes in that some molecular species of PG undergo a phase change above 0°C. The proportion of high-melting fatty acids is generally higher in chilling-susceptible plants than those that are not susceptible, but there is not the expected relationship between sensitivity and composition within a given susceptible plant species. The PG hypothesis is still an attractive one, particularly in light-enhanced chilling, as PG is important in maintaining stability of light-harvesting complexes. Some experimental evidence is accumulating that suggests that glucocerebrosides can induce similar phase separations in the plasma membrane and tonoplast as PG in thylakoids (Lynch, 1990).

Another compositional change in membranes during the development of chilling injury is the peroxidation of fatty acids (Parkin et al., 1989). Peroxidative changes are noted in most membrane-associated disorders of plants, but it has never been clearly established whether peroxidation represents a cause of injury or merely a secondary effect resulting from general degradation after the injury has become irreversible. It is well known that both hydrolysis and peroxidation both individually and in concert can induce biophysical changes within a membrane. A recent theory suggests that chilling injury results from inhibition of membrane-bound protein activity via changes in the physical properties of membrane lipids induced by peroxidation (Shewfelt and Erickson, 1991). The theory partitions peroxidation into initiation which could occur reversibly during injury at levels below normal detection and propagation which would not become evident until rewarming. Propagation, normally controlled by free-radical chain stoppers such as α-tocopherol, would represent the point between reversible and irreversible damage.

Sterols represent another lipid class in membranes with potential for modification and subsequent alteration of biophysical properties. Hugly et al. (1990) describe a chilling-susceptible mutant of *Arabidopsis* that shows a 10-fold increase in steryl esters during exposure at 13°C for 25 hr. Little change occurs in free sterols or neutral lipids. Chilling damage follows as evidenced by chlorosis and an increase in ion leakage. Guye (1988) reports an increase in the sitosterol to stigmasterol molar ratio in the least susceptible bean genotype studied and a decrease in the ratio of the most susceptible genotype during rewarming after chilling at 5°C for 24 hr. He postulates that a general increase of free sterols to phospholipids during chilling and the ratio of free sterols determines membrane response to chilling temperatures. Whitaker (1991) finds little change in phospholipid composition during 12 days storage of tomato fruits at 2°C. Total sterol composition increases during storage at 2°C and 15°C (slow ripening) with greater increases in free sterols at 2°C and steryl esters at 15°C. In addition the sitosterol to stigmasterol ratio remains constant at 2°C but decreases at 15°C. Free fatty acids accumulate during ripening (15°C) but not during chilling (2°C). These three studies present very different pictures of sterol accumulation in chilling-susceptible tissue during exposure to low temperatures which defy general conclusions.

While most compositional studies focus attention on the hydrophobic core of the membrane, surface characteristics are critical in membrane rigidification (Parkin et al., 1989). As the temperature decreases the structure of water near the surface would be expected to change affecting localized solute concentrations at the interface. Binding of two negatively-charged phospholipid molecules by a divalent cation such as Ca^{++} or Mg^{++} would serve to rigidify the membrane by restricting rotational motion. Little direct evidence is available linking chilling injury and surface properties of the membrane.

10.2.4. *Biophysical properties*

Much of the evidence for the biophysical changes in membranes associated with chilling injury has been derived by analogy. Early studies conclude that split Arrhenius plots are sufficient to establish phase transitions, but such a practice is now generally discredited (Steponkus, 1981; Lynch, 1990). Numerous studies report compositional changes in chilling-susceptible species at chilling exposures and speculate on the biophysical consequences based on analagous studies of changes observed during senescence or properties of liposomes with defined composition. The few studies that have actually investigated thermal properties of membranes as directly affected by chilling do not present a clear picture.

In the most comprehensive study to date Raison and Orr (1986) determine phase transitions for thylakoids isolated from mung bean, oleander and tomato leaves using differential scanning calorimetry (DSC), electron spin resonance (ESR) and fluorescence depolarization with *trans*-parinaric acid as a probe. DSC exotherms reveal transitions at 7, 10 and 12°C respectively for oleander, mung beans and tomato. ESR and *trans*-parinaric acid fluorescence suggest similar transition temperatures. These transition temperatures are all close to the critical temperature thresholds needed to induce injury. Murata and Yamaya (1984) note higher phase transitions in liposomes containing PG extracted from leaf membranes of chilling-susceptible species (sponge cucumber, squash and sweet potato) than from PG liposomes of non-susceptible species (cluster amaryllis, lettuce and spinach). Low et al. (1984), however, could detect no discernable pattern of DSC exotherms of thylakoids isolated from chilling susceptible species (cantaloupe, kidney bean, soybean and domestic tomato – *L. esculentum*) as compared with those from non-susceptible species (pea, spinach and wild tomato – *L. hirsutum*).

The studies presented above presume bulk changes in membrane lipids are required for changes of biophysical properties. Freeze-fracture evidence suggests that microdomains are formed during chilling of avocado fruit (Platt-Aloia and Thompson, 1987). No external symptoms of chilling injury were observed in fruit stored for 2 weeks at 6°C. Plasma membrane isolated from these fruit that had been rewarmed show a typical distribution of proteins, but plasma membrane isolated from fruit that had not been rewarmed show areas free of protein. Application of exogenous ethylene to chilled fruit induces chilling-injury symptoms within 2 weeks. Plasma membranes from the injured fruit, whether or not they are rewarmed, show similar protein-free domains. The authors conclude that the protein-free domains represent lateral phase separations which occur at an early, reversible step in the process that leads to irreversible injury.

Frenkel (1991) attributes chilling injury to increased molecular ordering of water leading to increased hydrogen bonding analagous to a dehydration response. This process would serve to destabilize membrane lipids and proteins by disrupting normal interaction with water. Membrane stabilization could be achieved by increasing hydrophilicity of the hydrophobic core. By this view, an increase of fatty-acid unsaturation and lipid peroxidation could serve as protective mechanisms from chilling injury by increasing hydrophilicity.

Additional postulated changes in biophysical properties due to chilling are based on studies with liposomes and other membrane-associated degradative processes as reviewed by Parkin et al. (1989) and Lynch (1990). Postulated biophysical changes include a decrease in fluidity with increased fatty-acid saturation, lamellar to hexagonal$_{II}$ phase transitions resulting in nonbilayer formations due to partial dehydration at the hydrophillic surface, rigidification of membranes by binding head groups with divalent cations due to altered water status at the surface, and broadening of phase transitions as a result of sterol accumulation within the membrane. Unfortunately little or no direct evidence linking these compositional changes to biophysical properties during chilling exists.

10.2.5. *Protein function*

Explicit in the original Lyons (1973) hypothesis and implicit in subsequent versions of the theory is that membrane dysfunction is a direct result of dysfunction of membrane-bound proteins. Although leakage of ions and small molecules are also considered consequences of the biophysical changes in the membranes and indicative of damage, protein dysfunction has received primary emphasis. Studies in three laboratories provide direct evidence for biophysical mediation of membrane-bound protein function in vitro in membranes isolated from plant tissue as reviewed in greater detail by Lynch (1990).

Biophysical changes in both lipids and proteins of smooth microsomal membranes of bean cotyledons are reported by Lynch et al. (1987). The pattern of changes with temperature in lipid fluidity as determined by diphenyl hexatriene (DPH) parallel conformational changes in membrane proteins as measured by the electronic paramagnetic resonance (EPR) probe 3-maleimido proxyl (3-MP). The patterns suggest that subtle changes in molecular ordering of the lipids rather than phase transitions are responsible for altered protein conformation.

Caldwell and Whitman (1987) note similar results in plasma membrane-enriched barley root microsomes. EPR spin labels show discontinuities in membrane lipids at 12°C corresponding to conformational changes in plasma membrane proteins also determined by EPR. In a companion study Caldwell (1987) shows that intrinsic fluorescence of plasma-membrane ATPase in these microsomes is also altered at 12°C. They propose a model suggesting that molecular ordering of lipid clusters around critical proteins such as ATPase within the membrane reduces the mobility of proteins thus restricting protein-protein interaction.

Li et al. (1990) compare the effects of intrinsic proteins on lipid mobility using ESR spin probes in thylakoids of leaves from chilling-susceptible (cucumber and tomato) and non-susceptible (lettuce and pea) plants. Lipids are more restricted in chilling-susceptible thylakoids than in those isolated from non-susceptible tissue. The results indicate that greater lipid-protein interaction occurs in thylakoids of chilling-susceptible species and that such interaction could affect protein function.

These studies provide the best direct link between physical properties of the membrane lipids and activity of membrane-bound proteins, but these results must be viewed with some caution. As Wise et al. (1990) demonstrate, it can be very dangerous to extrapolate from in vitro systems to in vivo. Conformational changes in proteins as suggested by discontinuities in EPR plots may not directly affect activity. In studies on *Anacystis nidulans* response to chilling exposures reviewed by Murata (1989) phase separations are not directly related to membrane dysfunction. In this review, Murata shows that breaks in Arrhenius plots for photosynthesis correspond to liquid-to-gel separation found in thylakoid membranes. Leakage of electrolytes and amino acids occur at temperatures just above the separations of plasma membranes. Irreversible damage to photosynthesis and the Hill reaction, however, is not noted until the temperature is lowered to that of the temperature of the plasma membrane phase separation. These results indicate either that irreversible damage to proteins occurs as a result of further changes in lipid physical properties below the phase separation temperature or that irreversible damage in the thylakoids is more closely related to modification of the plasma membrane.

Evidence presented in numerous studies support the contention that membrane lipid composition at the time of chilling exposure predetermines the biophysical response of the membrane and that these biophysical properties affect the cell in susceptible tissue via increased permeability and altered protein function. The critical chemical components, the biophysical mechanism and the specific membrane proteins responsible for chilling injury are still not clear. No single explanation may be sufficient for all cases.

10.3. Freezing injury

Freezing injury is frequently grouped with chilling injury under the guise of low temperature injury. Many researchers link the two syndromes together via a common mechanism. Plants susceptible to chilling injury tend to be at increased risk to freezing injury. Both variants of low temperature injury are characterized by loss of membrane permeability and altered metabolic activity. These effects are thought to be mediated by biophysical properties of the membrane as influenced by lipid composition. Investigators of both syndromes have been strongly influenced by the phase transition hypothesis of Lyons (1973) and its subsequent modifications as described in the previous section. Some changes in lipid composition, biophysical properties and altered enzyme activity are common to both chilling and freezing injuries. Extreme caution should be taken on drawing the analogy between the two injuries so closely that critical differences are obscured.

Freezing injury is frequently partitioned into that incurred above −5°C and injury at lower temperatures (Lindow, 1990). Species such as citrus fruits, beans and tomatoes (chilling susceptible) as well as corn and lettuce (not susceptible to chilling) are considered freezing susceptible because they are not able to tolerate intercellular ice formation. Other considerations include freeze dehydration at membrane interfaces and the ability to reform lipid bilayers after mechanical disruption by ice crystals. Freezing injury research currently focuses on three questions. What is the primary event in injury? What factors differentiate between freezing-susceptible and freezing-resistant tissues? What modifications can be made to decrease susceptibility in affected species?

10.3.1. *Intercellular localization*

The plasma membrane is generally considered the primary site of freezing injury. Injury can be incurred by freeze-induced cell dehydration due to efflux of intracellular water into extracellular spaces. Maintenance of membrane integrity upon thawing is critical to the survival of the cell. Intracellular ice is formed upon supercooling of the cytosol followed by ice nucleation within the cell. The plasma membrane regulates supercooling of the cytosol by regulating water flux through an alteration of membrane permeability. The plasma membrane also serves as a barrier between extracellular ice formation and cytosol by uncertain mechanism. An understanding of the role of the plasma membrane in preventing intracellular ice formation is critical to the understanding of freezing injury. Steponkus (1990) reviews the evidence that mechanical failure of the membrane may be the primary effect of freezing response that leads to intracellular nucleation.

Protoplasts isolated from rye (a cold-hardy tissue) have been used to study the ability of cells to withstand osmotic shock to simulate freeze-thaw cycles (Steponkus, 1990). Protoplasts are able to exclude membrane components by endocytotic vesiculation and survive the accompanying osmotic shock. There is a limit, however, to the amount of endocytotic vesiculation a protoplast can withstand before it undergoes expansion-induced lysis. Elasticity of the protoplast is estimated at 2–3%. Ultrastructural studies of freezing of plant cells substantiate the importance of the plasma membrane but question the relevance of protoplast studies to intact cells (Singh and Miller, 1985). Freezing results in vesiculation of the cell with the plasma membrane in a pattern distinctly different from protoplast response. Aggregation of intermembranous protein particles is observed in plasma membranes and other cellular membranes during freezing prior to vesiculation (Yoshida and Uemura, 1990) suggesting the formation of microdomains.

Although most freezing research focuses on the plasma membrane, some evidence implicates thylakoids. Steffen and Palta (1987) report reversible damage to photosynthetic processes in potato plants exposed to sublethal freeze-thaw regimes. Inhibition of photosynthesis occurs prior to any evidence of damage to the plasma membrane (ion leakage). Recovery of photosynthesis is noted during rewarming under low light conditions whlle high light enhances damage.

10.3.2. *Composition*

Many of the same type of compositional differences associated with chilling injury are postulated to affect freezing response. In freezing injury research, however, more emphasis is placed on altered water status at the membrane interface due to dehydration. Thus, lamellar-to-hexagonal phase transitions attract more interest in freezing injury studies. Molecular species of acyl chains and the potential role of peroxidation are also targets of investigation.

The distribution of phospholipid head groups within the plasma membrane affects lamellar-to-hexagonal phase transitions (Steponkus, 1990). Phosphatidylcholine (PC) readily forms bilayers and its presence in the plasma membrane helps maintain the lamellar phase. Phosphatidylethanolamine (PE), the other predominant phospholipid of the plasma membrane, is a non-bilayer forming lipid and enhances the transition to a hexagonal phase. See also Chapter 7. As the PE:PC ratio increases the potential for hexagonal phase (non-bilayer) formation upon dehydration increases.

Phase changes within the bilayer from liquid-crystal to solid-gel are also postulated to play a role in freezing injury. As the degree of unsaturation of acyl chains increases the phase transition temperature should decrease and the tissue should be more likely to withstand freezing temperatures. Since there are no consistent, clearly-distinguishable differences in lipid composition in studies comparing acclimated and non-acclimated tissue, differential response is attributed to differences in component molecular species. Of particular interest are those species of PC and PE containing unsaturated fatty acids at both *sn*-1 and *sn*-2 positions (Steponkus, 1990). Studies with isolated protoplasts show that fusion with liposomes containing PC molecular species consisting of one (16:0/18:1, 16:0/18:2) or both (18:1/18:1, 18:2/18:2, 18:3/18:3) unsaturated fatty acids increase resistance to freezing down to −5°C. Fusion with liposomes containing only saturated PC molecular species (14:0/14:0, 16:0/16:0) provide no protection.

Other lipid components may also modulate membrane fluidity in response to freezing stress (Steponkus, 1990). Plasma membranes that contain both free and glycosylated sterols are expected to have broad phase transitions. Molecular species of other phospholipids such as PG, phosphatidylserine (PS) and phosphainositol (PI) might form microdomains. The high-melting glucocerebrosides offer the greatest such potential in plant plasma membranes (Lynch, 1990).

Evidence for peroxidation within plant membranes is provided in microsomal membranes isolated from wheat plants exposed to ice-encasement stress at −1°C (Hetherington et al., 1987, 1988). Ice-encasement can lead to loss of plant viability without formation of intracellular ice. The microsomal fraction was selected for observation out of a concern that bouyant density separation of plasma membranes might not recover damaged membranes. Free fatty acid concentrations increase by 2 to 4 fold with little change in fatty acid unsaturation. The lack of accumulation of lysophospholipid suggest that phospholipase A_2 activity is not likely, but rather free fatty acid accumulation is due to either lipolytic acyl hydrolase activity or chemical deesterification by active oxygen species. The lack of change in fatty acid unsaturation suggests no lipid peroxidation occurs during injury, but a decrease of fatty acid unsaturation during the post-thaw recovery indicates that lipid peroxidation occurs as a secondary effect.

In a subsequent study McKersie et al. (1990) report that propagation of lipid peroxidation could be rapidly induced in liposomes prepared from soybean lecithin. Similar peroxidative challenges, however, could not induce propagation in wheat crown microsomes despite higher levels of unsaturated fatty acids. Thus, liposomes do not appear to be good models for studying peroxidation in plant membranes. One must also be careful in attributing peroxidative damage solely to secondary effects (Shewfelt and Erickson, 1991). Single-point initiation of peroxidation may occur in membranes during injury at levels not detectable by most procedures. Once formed each peroxidized fatty acid is able to propagate itself 100 fold prior to termination. Control of propagation is achieved primarily by the presence of antioxidants or repair pathways.

Maintenance of membrane integrity after exposure to low temperature stress appears to be a critical factor in limiting or reversing freeze damage. Both calcium ions (Ca^{2+}) and abscisic acid (ABA) could play a role in stabilizing membranes. Arora and Palta (1988) induce reversible freezing damage in onion bulb cells at −8.5°C and irreversible damage at −11.5°C. Losses of membrane-associated Ca^{2+} occur at both temperatures but loss of fluorescence intensity are two-fold greater at −11.5°C than at −8.5°C. Addition of $CaCl_2$ ameliorates freeze-thaw damage suggesting that binding of Ca^{2+} to phospholipid head groups helps stabilize the membrane. Chen and Gusta (1983) report that exogenous application of ABA enhances resistance to freezing. Although these authors ignore membranes in their explanation, ABA is known to stabilize membranes. Leshem et al. (1990) show that ABA can bind a saturated molecular species of PC (16:0/16:0). Such binding serves to rigidify membrane lipids protecting the membrane from bursting. By enhancing the mechanical properties of the membrane, ABA would help slow the potential for ice nucleation within the cell. Note that this requirement for increased saturation of fatty acids to stabilize the membrane is contrary to the general concept that unsaturated fatty acids are needed to maintain membrane fluidity.

10.3.3. *Biophysical properties*

Studies of biophysical response to freezing seek to characterize either the effects of temperature on the fluidity of the hydrophobic core or the effects of surface dehydration on head-group behavior. Fluidity studies employ fluorescence depolarization or X-ray diffraction. Head-group behavior is monitored by ESR.

Model phospholipid membranes are used to determine the theoretical effects of surface dehydration on gel-to-liquid transitions within the membrane (Steponkus, 1990). Molecular species (16:0/18:1, 16:0/18:2, 18:2/18:2, 18:3/18:3) of PC common to rye leaves are studied using DSC. Although an increase in phase transition temperatures by 40°C is observed as the water content decreases from 20 to 5% the effect is not sufficient to induce formation of solid-gel phase lipids at a dehydration equivalent of −10°C. Thus, DSC patterns do not support the formation of lateral phase separations during freezing.

Use of model lipid dispersions may not readily apply to plant membranes. Changes in fluorescence anistropy of DPHX reveal inflection points at −16°C and −8°C in plasma membrane preparations of acclimated and non-acclimated orchard grasses (Yoshida and Uemura, 1990). These inflection points closely correspond to temperatures of lethal freezing injury. Liposomes derived from the same preparations, however, do not show similar inflection points suggesting that membrane proteins are important contributors to phase behavior.

Internal modification of membrane lipids may also affect biophysical properties. Kendall et al. (1985) note differences in thermal of microsomes isolated from wheat crown seedlings lethally frozen at −12°C. X-ray diffraction patterns reveal a 40°C increase in gel-to-liquid transition temperature as a result of freezing injury. No differences are noted in fatty-acid unsaturation or free sterol content, but a 13-fold increase occurs in the free fatty acid to phospholipid ratio upon freezing. They postulate that free fatty acids are responsible for solid gel separations within the membrane.

The possible role of head group interactions in the development of freezing injury cannot be overlooked, particularly with respect to lamellar-to-hexagonal phase transitions. Although hard experimental evidence is lacking, Steponkus (1990) concludes that lateral phase separation leading to protein aggregation is more likely due to formation of hexagonal-phase rather than gel-phase lipids. Singh and Miller (1985) support this conclusion. Rye cells, labelled with the ESR probe 5NS (the N-oxyl-4, 4-dimethyloxazolidine derivative of 5-hexostearic acid) show increased molecular ordering during freezing. A more rapid ordering occurs in hardened than in nonhardened cells indicating that head-group induced disorder such as nonbilayer structures decrease resistance to freezing.

The role of proteins in biophysical behavior is also being studied. A trend in cold response biotechnological research is the transgenic transfer of *ice nucleating protein* (*inaZ*) from gram negative bacteria

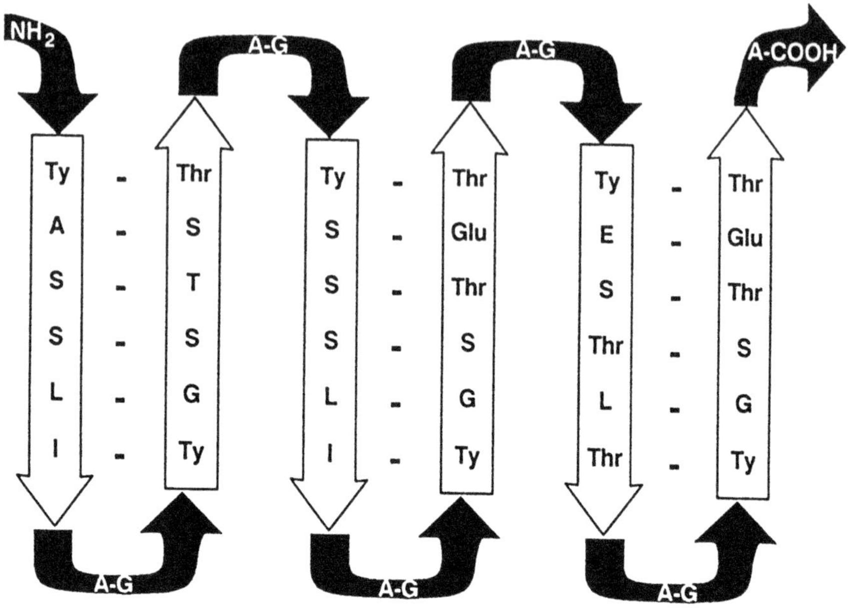

Fig. 10.3. The proposed structure of the ice-nucleating protein of *Pseudomonas syringae* as suggested by Lindow and Kajava (1991). Note that it is envisaged to contain 48 amino acids arranged in 6 arms (3 loops) each containing eight acids. A-G is invariably located on the hairpin bend sections. This protein moreover has several repetitive and conserved regions (Thr-S-G). The protein assumedly lies flat on the bacterial plasma membrane surface and is anchored on each extremity, respectively, by the NH_2 and COOH terminal groups. – indicates H^+ bonding. (This figure is reproduced by the courtesy of Dr. S.T. Lindow, Department of Plant Pathology, University of Berkeley, California.)

such as *Pseudomonas syringae*, *Erwinia* spp. and others, to higher plants. The introduction of such protein can considerably lower the supercooling capacity of host cell cytoplasma thus endowing the ability to withstand temperature stress which ordinarily would induce freezing. The $\Delta T°$ freezing point depression in this respect as found in transgenic species such as *Nicotiniana tabacum* and *Solanum* spp. varies between $-2°C$ to $-10°C$. Recent research conducted at the University of Berkeley (Lindow and Kajava, 1991) has indicated that the ice nucleating protein is located on the outer membrane of the bacterium and possesses 3 loops and 6 arms with the 48 amino acid sequence and configuration as shown in Fig. 10.3. This figure shows that the pair A-G is invariably located at the "hairpin bends" and that the protein possesses some highly conserved repetitive regions observed as Thr-S-G in certain of the arms.

As opposed to the ATPase type of multihairpin bent enzymes, this protein apparently *does not* span the bilayer but rather lies flat upon it. It has been suggested, but is yet not experimentally proven, that the *inaZ* can expand and contract like a concertina and its activation only at low temperatures is triggered by low-temperature induced compression. Because of the prokaryotic resemblance of the chloroplast membrane to the bacterial one major steps forward in ice nucleation protein transfer has been made between bacterial and chloroplast membranes. At present, commercially feasible transgenic ice-nucleated horticultural cultivars have not been produced.

10.3.4. *Protein function*

The enzyme most closely associated with freezing damage of plant tissue is plasma membrane H^+-ATPase (Yoshida and Uemura, 1990). It is a major protein in the plant plasma membrane comprising approximately 15% of the total protein present (Skulachev, 1988). It is distinguished from other cellular ATPases in that it is an E_1, E_2 type of ATPase signifying the formation of a phosphorylated intermediate and inhibition by vanadate. Its key role in growth and nutrition of the cell emphasize its vulnerability to membrane damage. Plasma membrane H^+ATPase is closely regulated by plant hormones and sensitive to external stress.

Ishwari and Palta (1991) follow the effects of freezing and thawing on plasma membrane H^+ATPase in potato leaves. Upon slight damage, as indicated by 10–20% water soaking and an increase in ion leakage from 15–27%, enzyme activity increases by 50% suggesting a potential role in injury recovery. As injury progresses, however, plasma membrane H^+ATPase activity declines with concomitant increases in water soaking and ion leakage and a decrease in leaf turgidity. These results are in agreement with previous studies (reviewed by Yoshida and Uemura, 1990), but are unique in that they suggest that the ATPase is involved early in the injury process. The results support membrane-lipid mediation of enzyme activity in response to stress but do not exclude other possibilities such as a cold-induced conformational change in the protein.

10.4. Cold acclimation

Certain plant species are able to decrease their susceptibility to low temperature stress during a modification of environmental conditions.

This process, known as cold acclimation, can greatly reduce low temperature injury. The two most intensively studied aspects of cold acclimation are cold-hardening of grains to reduce freezing injury and conditioning of detached fruits and vegetables to reduce postharvest chilling injury. In the original Lyons (1973) hypothesis, increased unsaturation of fatty acids during acclimation was offered as the explanation. Fatty acid unsaturation is still a major focus of acclimation research although head-group modification, resistance to peroxidation and the presence of membrane cryoprotectants are also being investigated.

The primary site of attack, thylakoids or plasma membrane for chilling injury and plasma membrane for freezing injury, is expected to be the primary site of acclimation. Some evidence suggests that injury may not be confined to a single membrane. Rather, organelles might interact in the etiology of injury and in acclimation. The most compelling evidence for such organellar interaction comes from chilling injury studies with *Anacystis nidulans* reviewed by Murata (1989). Although loss of photosynthetic capacity was identified as the primary indicator of irreversible damage, injury is incurred at temperatures corresponding to phase transitions in plasma membranes and not to those in thylakoids.

10.4.1. *Compositional changes*

During cold acclimation membrane lipids can be retailored to increase unsaturation of the fatty-acid chains (Thompson, 1989). Experimental evidence suggests that desaturases, which add double bonds to the fatty acids, are very selective in the molecular species they modify. The greatest potential for such retailoring to alter membrane phase properties is offered by the high-melting molecular species of PG in thylakoids and glucocerebrosides in plasma membrane. The best evidence for the importance of fatty-acid unsaturation in *A. nidulans* comes from a study by Vigh et al. (1985). Using a palladium catalyst they are able to increase fatty-acid unsaturation by adding hydrogen across the double bonds. In intact cells fatty acids in the outer membrane are hydrogenated first permitting preferential saturation of the plasma membrane. Irreversible inactivation of photosynthetic electron transport occurs as fatty acids of the plasma membrane become saturated.

Retailoring of fatty acids does not necessarily require formation of polyunsaturated fatty acids or the high-melting molecular species. Demandre et al. (1987) describe a relationship between the accumulation of relatively saturated molecular species of PC(16:0/18:1; 18:1/18:1) in apple fruits as they acquire resistance to freezing injury. These results correspond to studies with rye protoplast-liposome fusion reviewed by Steponkus (1990) and discussed in the previous section showing increasing resistance to freezing with the addition of PC containing one or two unsaturated fatty acids. Thus, the incorporation of the first double bond may be more important than additional unsaturation.

Other potential changes to the hydrophobic core during low temperature injury include hydrolysis or peroxidation of fatty acids. Loss of phospholipid and accumulation of free fatty acids is frequently reported during development of low temperature injury. McKersie et al. (1990) observe that microsomes from acclimated winter wheat crowns contain less saturated fatty acids and less free fatty acids than microsomes from non-acclimated crowns, but the acclimated microsomes are more susceptible to peroxidative attack. Although it is difficult to draw conclusions from a single study, particularly in the absense of molecular species data, the results argue against a direct role for hydrolysis or peroxidation during acclimation.

In addition to changes in acyl chains during acclimation, lipid classes are also affected (Steponkus, 1990; Yoshida and Uemura, 1990). In general phospholipids increase relative to sterols with free sterols increasing relative to steryl glucosides. More than half of the glucocerebrosides present in plasma membranes of winter rye decreases during acclimation. These changes are consistent with the hypothesis that membrane fluidity plays a role in freezing resistance.

Most views of membrane physiology related to low temperature injury and acclimation tend to assume a reasonably static process under "normal" conditions and attribute any modifications to direct cellular response to stress. Everything we know about the cell, however, suggests a dynamic physiological system. Steer (1988) estimates complete turnover of the plasma membrane in a plant cell is completed within 4 hours. Thus, the machinery is in place for rapid degradation and renewal of cell membranes, and rapid change is the rule not the exception. The precise mechanism of membrane turnover is not known but removal of "old" components is achieved by endocytosis and synthesis of "new" components occurs in the endoplasmic reticulum and Golgi apparatus. Compositional changes during development of reversible or irreversible injury should be evaluated in terms of membrane turnover.

10.4.2. *Biophysical changes*

Modification of lipid composition of membranes durlng acclimation is postulated to increase fluidity at low temperatures. The limited data collected on biophysical adaptation of plant membranes during acclimation are contradictory.

Studies of the response to *Anacystis nidulans* cells cultured at two temperatures (28°C and 38°C) provide strong support for the phase transition/fluidity hypothesis (Murata, 1989). Phase separations of plasma membranes from cells grown at both temperatures are reported using freeze-fracture electron microscopy, carotenoid absorption and ESR spin probes while phase separations of thylakoid membranes are reported using chlorophyll fluorescence, ESR spin probes and X-ray diffraction. Separations for plasma membrane (5°C) and thylakoids (13–16°C) are lower from cells grown at 28°C than separations obtained for plasma membrane (13–16°C) and thylakoids (23–26°C) from cells cultured at 38°C.

The data collected for thermotropic properties of membranes from higher plants as affected by acclimation is less clear (Yoshida and Uemura, 1990). Data using ESR spin probes suggest that protoplasts of winter wheat become more fluid during acclimation, but the specific incorporation of the probe was questionable. Little differences occur during acclimation in plasma membranes of orchard grass and Jerusalem artichokes using DPH as a fluorescence depolarization probe. Mulberry tree membranes show marked changes in the DPH anistropy parameter during the season (October-November) most closely associated with development of cold hardiness (Yoshida, 1984). These changes in membrane fluidity are attributed to a concomitant increase in phospholipid to sterol ratio.

Changes in bulk-phase phospholipids may not be necessary, however, if phase separations into microdomains within the membrane are responsible for injury. Yoshida and Uemura (1990) note that dramatic decreases in glucocerebrosides are reported during acclimation of winter rye whlle DSC profiles distinguish thermotropic properties of tonoplasts from mung-bean (chilling susceptible) and pea (non-susceptible) seedlings. No one has demonstrated, however, that the changes in glucocerebrosides during acclimation induce changes in phase behavior of plant membranes. Pike et al. (1990) observe decreases of two saturated molecular species (16:0/16:0; 18:0/18:0) of PG in cotton seedlings from 41% to 8% during acclimation at 15°C compared with normal growth at 35°C. Leaves from the acclimated seedlings evidence much less ion leakage during 6 days chilling in the dark at 3°C. No significant differences could be obtained for fluorescence depolarization of chlorophyll or critical temperature for fluorescence damage between the acclimated and non-acclimated tissue. It is not clear whether another physical measurement, such as DSC or ESR would reveal differences.

10.4.3. *Protein function*

If biophysical properties are modulating protein activity during acclimation, activity of target proteins should be higher in membranes of acclimated tissue exposed to low temperatures than non-acclimated tissue treated similarly. Some evidence exists to support this hypothesis in cold-hardening of grains.

Jian et al. (1982) compare the histochemical activity of plasma membrane ATPase in hardened and nonhardened seedlings of winter wheat subsequently held at 22°C and 5°C. Freezing tests confirm that the hardened seedlings could survive above −8°C while nonhardened seedlings are injured at −3°C. No qualitative differences are observed in plasma membrane ATPase activity at 22°C. At 5°C, however, hardened seedlings exhibit greater activity than nonhardened seedlings. The contrasting responses at the two temperatures support biophysical mediation of membrane lipids, but altered synthesis cannot be ruled out.

Huner et al. (1989) review a series of studies investigating the link between compositional and biophysical factors on photosynthesis during acclimation of winter rye. Both in vitro and in situ studies reveal enhanced photosynthetic capacity at low temperatures in hardened seedlings. Structure and function of PSI and LHCII are directly affected. A dramatic loss of the fatty acid trans-16:1 relative to 16:0 in molecular species of PG is noted with hardening contrary to postulated increase of fatty unsaturation described above. Freeze-fracture evidence suggests aggregation of proteins in membranes or nonhardened seedlings but not in those of hardened ones. These changes could not be linked, however, to any general changes in biophysical properties of the membrane lipids. They conclude that modification of membrane structure is responsible for the altered enzyme activity and suggest that specific lipid-protein interactions provide a key to understanding the acclimation process.

10.5. A concluding perspective

The hypotheses of Levitt (1962) and Lyons (1973) stimulated a myriad of studies on the role of membranes in the mechanisms of low temperature injury and cold acclimation. In the intervening decades the experimental data has tended to obscure rather than clarify our understanding of these processes. Despite the lack of resolution of important issues there are some common patterns emerging, and I offer the following generalizations.

Classification of both chilling and freezing as low temperature injury tends to obscure important differences in the two disorders. Freezing injury is more distinct, more closely associated with a physical event (intracellular ice formation), more reproducibly induced, and less diverse in symptom development than chilling injury. Even within each type of injury there tend to be important distinctions that become obscured. Light-induced and dark chilling injury may represent different types of response in different parts of the cell. Freezing injury induced just below 0°C appears to differ from that induced at much lower temperatures.

A single mechanism may not account for all the low temperature and acclimation responses. The temptation is strong to try to explain all responses in terms of a single unifying mechanism. One such mechanism may someday be revealed, but, in the interim, such efforts tend to thwart progress rather than enhance it. The bulk of experimental evidence argue against a unifying mechanism. While there may be some common steps in mechanistic pathways, there appear to be important differences in subcellular responses to chilling and freezing of photosynthetic and non-photosynthetic tissue of herbaceous and woody plants.

Lipid compositional studies provide little meaningful information on low temperature response. Compositional studies to date have demonstrated conclusively that an understanding of low temperature injuries or acclimation will not come solely from an understanding of the composition of bulk membrane lipids. Identification of molecular species associated with response to low temperatures may provide clues if coupled with specific changes in biophysical properties, loss of permeability or modification of enzyme activity. Likewise, studies linking sterol composition as affected by low temperatures to biophysical properties show promise.

Subtle changes in membrane composition and structure may be more important than dramatic ones. The bias of many approaches to the study of chilling and freezing injuries has been to explain differences by single dramatic changes. Unfortunately we are beginning to run out of dramatic changes to investigate, and we still do not have the answers we need. The weight of scientific evidence suggests that subtle changes in the membrane are responsible for the observed differences. Lipid-protein interactions within the membrane and the role of water at the membrane interface are two critical areas of study, that need more thorough investigation.

Membrane turnover is such an integral process within the cell that its relevance to chilling and freezing injury as well as to acclimation can no longer be ignored. Turnover is the process by which membrane composition is altered. Although changes in composition have been an integral part of the investigation of low temperature response, discussion has not been framed in a context of membrane turnover. The time required for injury development and acclimation is typically much longer than that of complete membrane turnover under normal circumstances. Thus, either chemical and physical changes observed during injury development or acclimation are relatively slow in a rapidly changing system or membrane turnover is dramatically slowed during injury development. If the former is the case, then the relevance of a primary event is questionable. If the latter is the case, the primary event is directly related to membrane turnover.

Isolated vesicles provide useful snapshots of chemical and physical properties of selected membranes within the cell but offer limited value in understanding dynamic response of plant cells to low temperature stress. Studies comparing vesicles isolated from rewarmed vs not rewarmed tissue or acclimated vs non-acclimated tissue emphasize the differences in in vitro and in vivo response. Caution must be exercised in using isolated vesicles to study low temperature response. The desire to study purified membrane fractions must be balanced with potential differences in partitioning due to changes in membranes induced by the stress conditions. Gross differences in membrane properties will be evident between vesicles from sample treatments but subtle differences such as lipid-protein interactions are much more difficult to determine. Extramembranal factors such as cell wall or cytoskeletal interactions are lost. Size of vesicles (e.g. large unilamellar vesicles vs small unilamellar vesicles) and sidedness can affect enzyme activity and must be considered in data interpretation. Despite these limitations, carefully-planned experiments can provide useful data on membrane properties. Isolated vesicles are also useful models for observing effects of specific compositional modifications on biophysical properties and enzyme activity.

Liposomes offer little or no use as models for low temperature response. The qualitative differences in behavior between liposomes and isolated vesicles limit their usefulness as models. Results of comparisons between isolated vesicles and liposomes emphasize the contribution of proteins to the biophysical properties of the membrane. Liposomes have been successfully employed in modification of lipid composition of vesicles or protoplasts via fusion to determine effects on physical properties and can provide a useful model for understanding the basic biophysical properties of lipids in bilayers.

Biophysical measurements must go beyond phase transitions and fluidity. Whether by fluorescence depolarization, DSC, X-ray diffraction, EPR, ESR or freeze-fracture evidence, biophysical studies have established qualitative differences, referred to nonspecifically as fluidity, in low temperature response. Little is known about the specific physiochemical parameters (e.g. chain ordering, lateral and rotational diffusion) as they are affected by low temperatures. Unfortunately, vesicles are not good models to study these parameters and liposomes are not good models for low temperature response. Innovative applications of existing techniques will be needed to link lipid composition to biophysical properties to direct effects on permeability or enzyme activity. A series of techniques, measurement of mechanical properties (e.g. surface pressure/molecular area isotherms and membrane collapse points), that can be used to estimate basic physiochemical parameters has been generally ignored in attempting to understand low temperature response.

Scientific evidence provides strong support for biophysical mediation of membranes in the development of chilling and freezing injuries as well as cold acclimation. The prevailing theory proposes that lipid composition determines biophysical response to low temperatures which in turn affects membrane permeability and enzyme activity. Despite numerous studies the specific mechanisms continue to elude investigators. To establish this theory and elucidate specific mechanisms, efforts must be directed to provide a clearer understanding of specific physiochemical parameters to low temperature response and providing better linkages between biophysical properties and both lipid composition and enzyme activity.

10.6. References and recommended reading

Arora, R. and Palta, J.P. 1988. *In vivo* perturbation of membrane-associated calcium by freeze-thaw stress in onion-bulb cells. *Plant Physiol.* 87:622–628.

Caldwell, C.R. 1987. Temperature-induced protein conformational changes in barley root plasma membrane-enriched microsomes. II. Intrinsic protein fluorescence. *Plant Physiol.* 84:924–929.

Caldwell, C.R. and Whitman, C.E. 1987. Temperature-induced protein conformational changes in barley root plasma membrane-enriched microsomes. I. Effect of temperature on membrane protein and lipid mobility. *Plant Physiol.* 84:918–923.

Chen, T.H.H. and Gusta, L.V. 1983. Abscisic acid-induced freezing resistance in cultured plant cells. *Plant Physiol.* 73:71–75.

Cheng, T.-S. and Shewfelt, R.L. 1988. Effect of chilling exposure of tomatoes during subsequent ripening. *J. Food Sci.* 53:1160–1162.

Demandre, C., Justin, A.M., Nguyen, X.V., Gawer, M., Tremolieres, A. and Mazliak, P. 1987. Molecular species of phophatidylcholine in plants. In Eds. P.K. Stumpf, J.B. Mudd and W.D. Nes. *The Metabolism, Structure, and Function of Plant Lipids.* pp. 273–282. Plenum Publ. Corp.

Frenkel, C. 1991. Disruption of macromolecular hydration – a possible origin of chilling destabilization of biopolymers. *Trends Food Sci. Technol.* 2(2):39–41.

Guye, M.G. 1988. Sterol composition in relation to chill-sensitivity in *Phaseolus* spp. *J. Exp. Bot.* 39: 1091–1096.

Hetherington, P.R., Broughton, H.L. and McKersie, B.D. 1988. Ice-encasement injury to microsomal membranes isolated from winter wheat crowns. II. Changes in membrane lipids during ice encasement. *Plant Physiol.* 86:740–743.

Hetherington, P.R., McKersie, B.D. and Borochov, A. 1987. Ice-encasement injury to microsomal membranes from winter wheat crowns. I. Comparison of membrane-properties after lethal ice encasement and during a post-thaw period. *Plant Physiol.* 85:1068–1072.

Hugly, S., McCourt, P., Browse, J., Patterson, G.W. and Somerville, C. 1990. A chilling sensitive mutant of *Arabidopsis* with altered steryl-ester metabolism. *Plant Physiol.* 93:1053–1062.

Huner, N.P.A., Krol, M., Williams, J.P., Maissan, E. and Krupa, Z. 1989. Development at cold-hardening temperatures: Membrane assembly and organization. In: Ed. P.H. Li. *Low Temperature Stress Physiology in Crops.* pp. 53–65. CRC Press. Boca Raton, FL.

Ishwari, S. and Palta, J.P. 1991. Plasma membrane ATPase as a key site of alteration by a freeze-thaw stress: Role in recovery and progressive injury. In: Ed. J.H. Cherry. *Environmental Stress in Plants.* pp. 259–265. Springer-Verlag, Berlin.

Jian, L-C., Sun, L-H. and Dong, H-Z. 1982. Adaptive changes in ATPase activity in the cells of winter wheat seedlings during cold hardening. *Plant Physiol.* 70:127–131.

Kendall, E.J., McKersie, B.D. and Stinson, R.H. 1985. Phase properties of membranes after freezing in winter wheat. *Can. J. Bot.* 63:2274–2277.

Leshem, Y.Y., Cojacaru, M., Margel, S., El-Ani, D. and Landau, E.M. 1990. A biophysical study of abscisic acid interaction with membrane phospholipid components. *New Phytol.* 116:487–498.

Levitt, J. 1962. A sulfhydryl-disulfide hypothesis of frost injury and resistance in plants. *J. Theor. Biol.* 3:355–391.

Levitt, J. 1972. *Responses of Plants to Environmental Stresses.* Academic Press, New York.

Levitt, J. 1980. *Responses of Plants to Environmental Stresses I. Chilling, Freezing and High Temperature Stresses*. Academic Press, New York.

Li, G., Knowles, P.F., Murphy, D.J. and Marsh, D. 1990. Lipid-protein interactions in thylakoid membranes of chilling-resistant and -sensitive plants studied by spin label electron spin resonance spectroscopy. *J. Biol. Chem.* 265:16867–16872.

Lindow, S.E. 1990. Use of genetically altered bacteria to achieve plant frost control. In: Eds. J.P. Nakas and C. Hagedorn. *Biotechnology of Plant-Microbe Interactions*. pp. 85–110. McGraw Hill, New York.

Lindow, S.E. and Kajava, A. 1991. Personal communication.

Low, P.S., Ort, D.R., Cramer, W.A., Whitmarsh, J. and Martin, B. 1984. Search for an endotherm in chloroplast lamellar membranes associated with chilling-inhibition of photosynthesis. *Arch. Biochem. Biophys.* 231:336–344.

Lynch, D.V. 1990. Chilling injury in plants: the relevance of membrane lipids. In: Ed. F.R. Ketterman. *Environmental Injury to Plants*. pp. 17–34. Academic Press, New York.

Lynch, D.V., Lepcock, J.R. and Thompson, J.E. 1987. Temperature-induced changes in lipid fluidity alter the conformation of proteins in senescing plant membranes. *Plant Cell. Physiol.* 28:787–798.

Lyons, J.M. 1973. Chilling injury in plants. *Ann. Rev. Plant Physiol.* 24:445–466.

Lyons, J.M., Graham, D. and Raison, J.K. 1979. *Low Temperature Stress in Crop Plants – The Role of the Membrane*. Academic Press, New York.

Marangoni, A.G., Smith, A.K., Yada, R.Y. and Stanley, D.W. 1989. Ultrastructural changes associated with chilling injury in mature-green tomato fruit. *J. Amer. Soc. Hort. Sci.* 114:958–962.

McKersie, B.D., Hoekstra, F.A. and Krieg, L.C. 1990. Differences in the susceptibility of plant membrane lipids to peroxidation. *Biochim. Biophys. Acta* 1030:119–126.

Minorsky, P.V. 1985. An heuristic hypothesis of chilling injury in plants: a role for calcium as the primary physiological transducer. *Plant, Cell Environ.* 8:75–94.

Minorsky, P.V. 1989. Temperature sensing by plants: a review and hypothesis. *Plant, Cell Environ.* 12:119–135.

Murata, N. 1989. Low-temperature effects on cyanobacterial membranes. *J. Bioen. Biomem.* 21:61–75.

Murata, N., Sato, N., Takahashi, N. and Hamazaki, T. 1982. Compositions and positional distributions of fatty acids in phospholipids from leaves of chilling-sensitive and chilling-resistant plants. *Plant Cell Physiol.* 23:1071–1079.

Murata, N. and Yamaya, J. 1984. Temperature-dependent phase behavior of phosphatidylglycerols from chilling-sensitive and chilling-resistant plants. *Plant Physiol.* 74:1016–1024.

Parkin, K.L., Marangoni, A., Jackman, R., Yada, R. and Stanley, D. 1989. Chilling injury. A review of possible mechanisms. *J. Food Biochem.* 13:127–153.

Pike, C.S., Norman, H.A., Kennerer, E.C., Wessner, D.R., Greenberg, C.M., Kaplan, L.J., Brodsky, N.M. and Ellis, A.A. 1990. Effects of acclimation to low temperature and to water stress on photosynthesis and on physical and chemical properties from thylakoids of cucumber and cotton. *Plant Sci.* 68:189–196.

Platt-Aloia, K.A. and Thompson, W.W. 1987. Freeze fracture evidence for lateral phase separations in the plasmalemma of chill-injured avocado fruit. *Protoplasma* 136:71–80.

Raison, J.K. and Lyons, J.M. 1986. Chilling injury: a plea for uniform terminology. *Plant Cell Environ.* 9:685–686.

Raison, J.K. and Orr, G.R. 1986. Phase transitions in thylakoid polar lipids of chilling-sensitive plants. *Plant Physiol.* 80:638–645.

Shewfelt, R.L. and Erickson, M.E. 1991. Role of lipid peroxidation in the mechanism of membrane-associated disorders in edible plant tissue. *Trends in Food Sci. Technol.* 2:152–4.

Singh, B.R. and Singhal, G.S. 1989. Role of surface proteins in the manifestation of temperature effect on thykaloid membranes. *Biochem. Physiol. Pflanzen* 185:1–10.

Singh, J. and Miller, R.W. 1985. Biophysical and ultrastructural studies of membrane alterations in plant cells during extracellular freezing: Molecular mechanism of membrane injury. In: Ed. K.K. Kartha. *Cryopreservation of Plant Cells and Organs*. pp. 61–73. CRC Press, Boca Raton, FL.

Skulachev, V.P. 1988. *Membrane Bioenergetics*. Springer-Verlag, Berlin.

Steer, M.W. 1988. Plasma membrane turnover in plant cells. *J. Exp. Bot.* 39:987–996.

Steffen, K.L. and Palta, J.P. 1987. Photosynthesis as a key process in plant response to low temperatures, alteration during low temperature acclimation and impairment during freezer-thaw injury. In: Ed. P.H. Li. *Plant Cold Hardiness*. pp. 66–99. Alan R. Liss, Inc., New York.

Steponkus, P.L. 1981. Responses to extreme temperatures. Cellular and sub-cellular bases. pp. 372–402. In: Eds: O.L. Lang, P.S. Nobel, C.B. Osmond and H. Ziegler. *Encyclopedia of Plant Physiology New Series* 12*A*: *Physiological Plant Ecology I*. Springer Verlag, New York.

Steponkus, P.L. 1990. Cold acclimation and freezing injury from a perspective of the plasma membrane. In: Ed. F.R. Ketterman. *Environmental Injury to Plants*. pp. 1–16. Academic Press, New York.

Thompson, G.A. 1989. Molecular changes in membrane lipids during cold stress. In: Ed. J.H. Cherry. *Environmental Stress in Plants*. Springer Verlag, Berlin.

Vigh, L., Gambos, Z. and Joo, F. 1985. Selective modification of sytoplasmic membrane fluidity by catalytic hydrogenation provides evidence on its primary role in chilling susceptibility of the blue-green alga, *Anacystis nidulans*. *FEBS Lett.* 191:200–204.

Whitaker, B.D. 1991. Changes in lipids of tomato fruit stored at chilling and non-chilling temperatures. *Phytochemistry* 30:757–761.

Wise, R.R., Terashima, I. and Ort, D.R. 1990. The effect of chilling in the light on photophosphorylation. *Photosyn. Res.* 25:137–139.

Yoshida, S. 1984. Chemical and biophysical changes in the plasma membrane during cold acclimation of mulberry bark cells (*Morus bombycis* Koidz. cv Goroji). *Plant Phys.* 76:257–65.

Yoshida, S. and Uemura, M. 1990. Responses of the plasma membrane to cold acclimation and freezing stress. In: Eds. C. Larsson and C.M. Moller. *The Plant Plasma Membrane*. pp. 293–319. Springer Verlag, Berlin.

11. The plasma membrane and tonoplast of guard cells

C.M. WILLMER and O. PANTOJA

11.1. Introduction

The forces needed to open stomata are osmotically generated by accumulation of K^+ by guard cells and concurrent uptake of external Cl^- and/or intracellular synthesis of malate to maintain electroneutrality. Potassium, Cl^- and malate collect in vacuoles of guard cells during opening and exit during closure. Environmental factors and some hormones stimulate and modify the ion fluxes thereby influencing stomatal movements. Thus ion transport across the tonoplast and plasma membrane play a central role in stomatal functioning.

Another feature of guard cells which makes them unique amongst higher plant cells is the very large change of cell surface area and volume that occurs with stomatal opening and closing. These changes of surface area may involve an unusually large turnover of membrane material of both the tonoplast and the plasma membrane.

In this chapter we review recent work on the above topics.

11.2. Morphological changes of the plasma membrane and tonoplast during guard cell volume changes

When stomata of *Commelina communis* open to about 12 μm aperture from the closed position, the outer dimensions (i.e., the perimeter of the cell wall) increase by about 10% (Willmer, unpublished data). The plasma membrane surface area, however, must increase by an even greater amount since, during guard cell swelling, the wall volume remains constant and must therefore be stretched in a thinner layer over the surface of the protoplasts. This extra volume and surface area is taken up by additional swelling of the protoplasts.

Plasma membranes have the physical ability to stretch by about 2–3% (Wolfe and Steponkus, 1983). During stomatal opening the additional increase of plasma membrane surface area must therefore be due to exocytotic incorporation of new membrane material or extension of infolds of existing plasma membrane. Upon closure the plasma membrane must become convoluted or endocytotic reabsorption of membrane material must occur.

Electron micrographs of guard cells often indicate that the plasma membrane has an undulating topography as it follows the inner surface of the cell wall. Thus, as stomata open and the guard cell wall is stretched and becomes thinner, the "slack" of the plasma membrane will be pulled to its full extensibility. Upon stomatal closure presumably the plasma membrane assumes its corrugated surface appearance. However, it is improbable that the unfolding of corrugations of the plasma membrane can account for all the surface area increase that occurs as stomata open fully. If this is the case then there must be a very active recycling of membrane within guard cells.

Coated pits (see pg 84 in Chapter 5) and vesicles are associated with endocytosis and have been observed in guard cell protoplasts (GCP) (Doohan and Palevitz, 1980). This suggests that endocytosis occurs in these protoplasts as they contract. However, we have never seen coated pits or vesicles in intact guard cells (with a cell wall) and are unaware of other published studies showing such structures in electron micrographs of intact guard cells. Indeed, there is a view that endocytosis only occurs in cells that have very low turgor (Gradmann and Robinson, 1989) and hence in plasmolysed cells such as protoplasts. Furthermore, our efforts to observe incorporation of vesicles of lucifer yellow into guard cells during stomatal closure, which would indicate endocytosis, have been unsuccessful (O'Parka and Willmer, unpublished data).

Thus, it is apparent that more investigations are needed to clarify the situation about possible membrane recycling linked to the plasma membrane. Even less is known about changes of shape and structure of the tonoplast during stomatal opening and closing. According to Guyot and Humbert (1970) guard cells of *Anemia rotundifolia* possess many small vacuoles when stomata are closed and fewer, larger ones when stomata open. Palevitz and O'Kane (1981), however, found that in differentiating guard cells of *Allium* the vacuole switches from a globular form to a network of interconnected tubules and then back to a globular form (with globules connected) at maturity. They did not investigate if the vacuole changed its form during stomatal movements. In another electron microscopy study (S. Smith and J.D.B. Weyers, unpublished data) in which thin cross-sections of an entire guard cell from a closed stoma of *Commelina* were made, again, small vacuoles appeared connected to a large central vacuole. Of further interest in this study was the finding that the surface area of the tonoplast was considerably greater than that of the plasma membrane.

From these descriptions of the shape of the vacuole it is unclear if there is a major insertion and extraction of tonoplast membrane during stomatal opening and closing even if the membrane topography does change during stomatal movements.

11.3. Energization of the plasma membrane of guard cells

Before studies of the biochemistry of guard cell membranes were made investigators had established that guard cells acidified incubation media inferring that there was a H^+ efflux from the cells. Initial studies used epidermal strips from leaves but, later, GCP were used and one recent study has investigated proton pumping in vesicles from GCP.

Electrophysiological investigations were also made in efforts to learn more about the energization of the plasma membrane.

11.3.1. *Proton pumping in epidermal strips*

Raschke and Humble (1973) first reported on medium acidification when stomata in epidermal strips of *Vicia faba* were induced to open. Since then, numerous studies have been made using a variety of methods for monitoring acidification (e.g., Gepstein et al., 1982/3; Bowling and Edwards, 1984; Edwards et al., 1988; Inoue and Katoh, 1987). Fig. 11.1A,B,C, shows results Fricker and Willmer (1990b) obtained when epidermal strips of *Commelina* were floated on a weakly buffered medium containing 50 mM KCl and stomata were induced to open in light and CO_2-free air. In darkness an initial acidification of the medium occurred which stabilised after about 1 h. This initial acidification is probably due to exchange of K^+ for H^- in the cell wall and to some stomatal opening that occurs even in darkness in the presence of exogenous KCl. Net H^+ efflux continues for up to 200 min and is accompanied by stomatal opening (Fig. 11.1A,B). A similar pattern of H^+ efflux occurs whether "intact" epidermis (epidermis with most cell types intact) or "isolated" epidermis (acid-treated epidermis which leaves only the guard cells intact – Squire and Mansfield, 1972) is used, though the level of acidification is re-

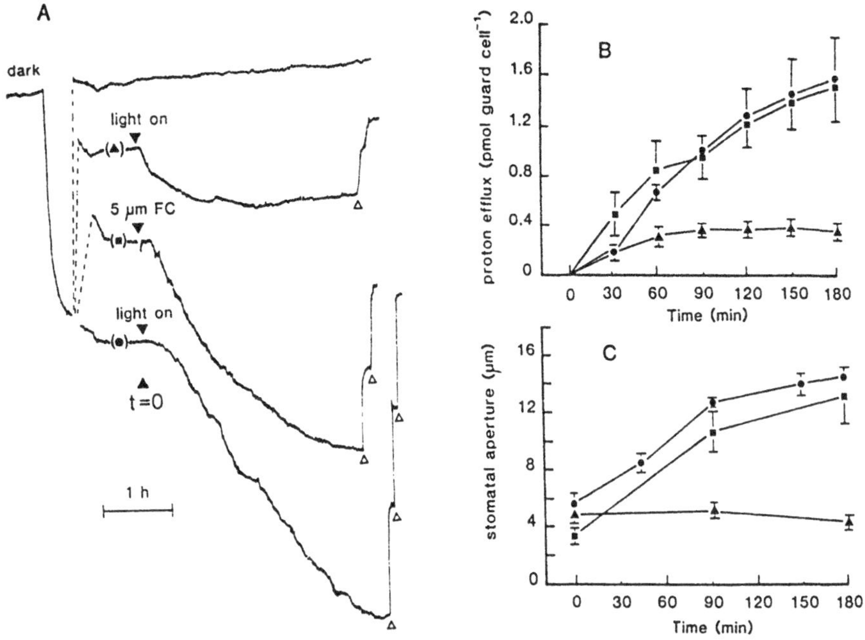

Fig. 11.1. Proton efflux and stomatal opening in epidermal strips of *Commelina communis*. *A*: Representative, pH traces from individual experiments in which, initially, epidermis was incubated in the dark at 25°C under CO_2-free conditions. When the pH of the incubating medium stabilised, epidermis was irradiated (200 μmol m^{-2} s^{-1} at 400–700 nm) at the points indicated in the absence or presence of vanadate. Fusicoccin (FC)-treated epidermis was maintained in darkness at a low KCl concentration. pH changes were calibrated against aliquots of NaOH (1 mM) as indicated. *B*: Proton efflux calculated after calibration of pH changes. *C*: Stomatal apertures measured in parallel experiments. Key: ●, light alone; ▲, light + vanadate (0.5 mM); ■, dark + FC (5 μM).

duced. Total H^+ efflux over an aperture increase from 5.6 μm to 15 μm was 0.64 pmol guard $cell^{-1}$ in "isolated" epidermis and, after a slight lag period of a few min, maximum H^+ efflux occurred over the next 30 min (4.1 fmol guard $cell^{-1}$ min^{-1}, equivalent to 36.3 nmol s^{-1} m^{-2} plasma membrane surface area). Such H^+ effluxes are similar to reported fluxes for K^+ uptake by guard cells suggesting a close connection between the two processes.

Fusicoccin stimulates stomatal opening (e.g., Squire and Mansfield, 1974) possibly by stimulating the plasma membrane ATPase via a receptor which binds fusicoccin (De Michelis et al., 1989). It is not surprising, therefore, to find that fusicoccin stimulates H^+ efflux which is associated with stomatal opening, even in darkness and at a low KCl concentration (Fig. 11.1A,B,C).

After a lag phase, vanadate (an inhibitor of plasma membrane

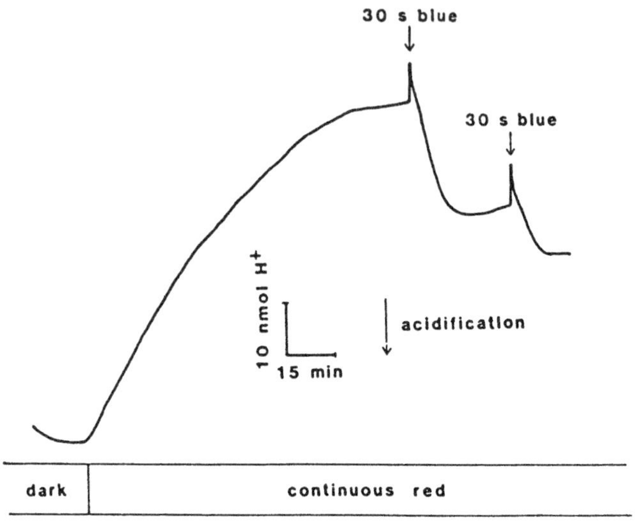

Fig. 11.2. Stimulation of proton efflux from guard cell protoplasts of *Commelina communis* (from Pantoja, 1988). Protoplasts were irradiated with continuous red (>600 nm) light (800 μmol m^{-2} s^{-1}) and 30 s pulses of blue (<500 nm) light (150 μmol m^{-2} s^{-1}) were superimposed at the points indicated.

ATPase) inhibited H^+ efflux and stomatal opening (Fig. 11.1A,B,C). However, variable results with vanadate have been obtained by different investigators. For example, Karlsson and Schwartz (1988) observed that vanadate did not inhibit opening in *Commelina* epidermis while Gepstein et al. (1982/83) reported that it did in *Vicia* epidermis. The effectiveness of vanadate is probably related to its rate of entry into the cells and we find that preincubation of the tissue for about 0.5 h at pH values below 6.5 gives consistent results.

11.3.2. *Proton pumping by guard cell protoplasts*

Medium acidification by GCP is stiulated by white light (Gotow et al., 1985) but measurement of H^+ efflux is complicated by CO_2 uptake causing alkalinization of the medium. Hence, H^+ efflux has been measured in CO_2-free air or in high photon flux densities (PFD) of red light which saturates photosynthesis. Under such conditions, pulses of low photon flux densities of blue light stimulate H^+ efflux by GCP of *V. faba* (Shimazaki et al., 1986; Shimazaki and Zeiger, 1987) and *Commelina* (Pantoja, 1988) (see Fig. 11.2). Pantoja (1988), however, found that vanadate caused a 30% inhibition of the blue-light induced H^+ efflux while Shimazaki et al. (1986) observed no effect of vanadate. Such apparently conflicting data is probably due to variable uptake of

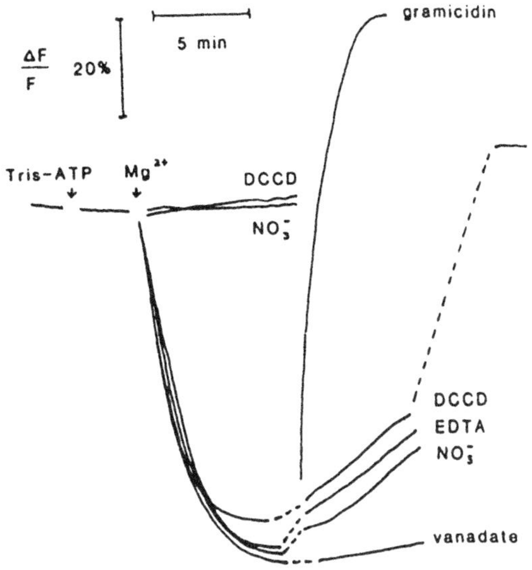

Fig. 11.3. Proton pumping in vesicles from guard cell protoplasts of *Commelina communis* as indicated by quenching of quinacrine fluorescence (from Fricker and Willmer, 1990a). Typical traces of Mg-ATP dependent quenching of quinacrine fluorescence in crude protoplast homogenates incubated at pH 8.0 are shown, together with the effects of various inhibitors of proton pumping.

the vanadate (see section 3.1). We consider that at least part of the uninhibited portion of the H^+ efflux may be due to a redox system located in the plasma membrane (but see Pantoja and Willmer, 1988, 1991).

11.3.3. *Proton pumping by vesicles from guard cell protoplasts*

One study has investigated the transport properties of vesicle preparations from GCP, namely that of Fricker and Willmer (1990a) (Fig. 11.3). Using a mixed vesicular extract from GCP of *Commelina* and fluorescent quenching of quinacrine at pH 8.0 it was found that formation of H^+ gradients were dependent on the presence of Mg^{2+} but inhibited by NO_3^-, EDTA and DCCD. Gramicidin, a H^+ ionophore, also collapsed the gradient. These characteristics indicate that the H^+ gradients are due to tonoplast ATPase activity (NO_3^-, specifically inhibits this ATPase which also has a pH optimum of 8.0 – see section 4). At pH 6.6 (the optimum for plasma membrane ATPase activity) no H^+ pumping was detectable. Possible reasons for this are that plasma membrane vesicles formed right-side out or leaky vesicles were formed unable to sustain a pH gradient.

At pH 8.0, K^+ had no effect on the formation of H^+ gradients, on the steady state level, or on the collapse of H^+ gradients. Thus, a H^+/K^+ antiporter system, which has been suggested to occur at the tonoplast of guard cells, was not indicated by the above criteria.

11.3.4. *Guard cell membrane potentials*

The observation that light and dark periods caused hyperpolarization and depolarization, respectively, of the guard cell plasma membrane was preliminary evidence of an active transport mechanism (Gunar et al., 1975; Zeiger et al., 1977). Zeiger et al., (1977) postulated that a photoactivated pump was responsible for the changes of membrane potential but it was not until the work of Assmann et al. (1985) that good evidence for this view was obtained. They employed the whole-cell mode of the patch-clamp technique to GCP and found that a 30 s pulse of low PFD blue light superimposed on high PFD red light induced a transient membrane hyperpolarization (also see section 11.3.2). However, if ATP was omitted from the pipette-filling solution, hyperpolarization did not occur indicating that the blue light effect was due to an electrogenic ATPase. Nevertheless, the outward current of the pump of 5.5 pA obtained in the Assmann et al. (1985) study is not enough to mediate K^+ uptake. In a later study (Schroeder, 1988), however, using slow-whole-cell recordings in which wash-out of the cytoplasm was prevented, blue light stimulated outward currents of up to 18 pA. Thus, in addition to Mg-ATP, the presence of other cytoplasmic components may be required to obtain optimal blue-light-stimulated pump activity. Schroeder (1988) also found that blue light hyperpolarized the plasma membrane to potentials more negative than -120 mV, values that should activate K^+ channels (see below).

Activation of the plasma membrane H^+-ATPase by light is not restricted to the blue region of the spectrum and high fluence rates of red light have also been found to stimulate outward currents in whole GCP (Serrano et al., 1988).

Stomata are also stimulated to open in intact leaves by red and particularly blue light. Thus, light-induced stomatal opening appears to be initiated, by stimulation of a H^+-ATPase which establishes a proton motive force (pmf) across the plasma membrane which, in turn, energises K^+ uptake by guard cells.

Blatt (1987), however, does not readily accept this view. Using a two-electrode voltage clamp method with intact guard cells he found a voltage dependant electrogenic pump with a stoichiometry of one positive charge/1 ATP. He argues that activation of the pump in the presence of a constant K^+ leak is insufficient to establish a voltage change that can drive K^+ uptake (membrane potential – potassium equilibrium potential) due to a decrease in pump activity at negative membrane potentials.

Few electrophysiological studies have been made of the guard cell tonoplast. However, patch-clamp experiments have confirmed the electrogenicity of the tonoplast ATPase. Using whole guard cell vacuoles and under current-clamp conditions with a *pH* gradient of 2 units (acidic inside) addition of ATP to the medium (cytoplasmic side) established a potential of around 40 mV (inside positive) (Raschke et al., 1988). Hence the H^+-ATPase is able to move positive charge against relatively large concentration gradients of H^+.

The observation that the H^+-ATPase activity of the tonoplast is modulated by light (see section 5.2) will also add new impetus to the electrophysiological studies.

11.4. Ion transport by guard cells

Activity of the membrane-bound H^+-ATPase 'pumps' generates an electrochemical difference for protons (the pmf) which can be used to drive the secondary transport of ions through channels or by porter systems.

11.4.1. *Ion uptake by guard cells*

Early studies established that guard cells accumulate large amounts of K^+ during stomatal opening (e.g., Imamura, 1943; Fujino, 1967). Later, numerous investigators used electrophysiological techniques to study the mechanism of this K^+ uptake but it was not until the patch-clamp work of Schroeder et al. (1984) that the existence of K^+ channels in the plasma membrane of guard cells was demonstrated. By clamping guard cell protoplasts to membrane potentials more negative than -100 mV it has been shown that inwardly directed channels specific for K^+ are activated (Schroeder et al., 1987; Schroeder, 1988). Thus, plasma membrane potentials more negative than -100 mV which can be obtained by stimulation of the H^+-ATPase by red or blue light could be used to drive uptake of K^+ for stomatal opening.

Uptake of K^+ by guard cells must be balanced by either uptake of an anion or by the synthesis of organic anions in order to maintain electroneutrality. Evidence indicates that a combination of the two processes is involved in most species but in the *Allium* genus Cl^- may be the major anion balancing K^+ (Schnabl and Raschke, 1980). Because of thermodynamic impositions (μCl^- cyt $\gg$ μCl^- outside), Cl^- uptake must be an energy consuming process which could be mediated by either a Cl^-/H^+ symport or a Cl^-/OH antiport.

Less is known about the movement of ions across the tonoplast of guard cells. Hedrich et al. (1988) detected a channel in the tonoplast that selected poorly between cations and anions, was activated by micromolar cytoplasmic calcium and by negative vacuolar potentials. Activation by negative tonoplast potentials of inward currents indicates either that cations move into the vacuole or that anions move out of the vacuole. However, since the physiological polarity of the tonoplast potential is positive relative to the cytoplasm, a negatively activated channel would not support K^+ uptake through such a channel. Presumably, therefore, K^+ enters the vacuole via a K^+/H^+ antiport system (but see section 3.3). Although Cl^- can move through the channel into the vacuole, this, again, seems inappropriate since the potential remains positive relative to the inside of the vacuole and a secondary active mechanism, probably driven by the pmf, must be postulated.

11.4.2. *Ion release by guard cells*

Closing of stomata is a consequence of the decrease in turgor pressure of guard cells caused by the loss of the solutes, K^-, Cl^- and malate. Perhaps the simplest explanation of the ion release is that, upon decreased activity of the H^+ pumps at the tonoplast and plasma membrane, the pmf would also decrease across both membranes and, consequently, the secondary active transport and passive transport dependent on hyperpolarization would halt. Of course, ions already in the guard cell must be allowed to exit and some observations can account for this. Schroeder et al. (1987) detected an outwardly conducting channel specific for K^+ which was stimulated by membrane potentials positive to the equilibrium potential for K^+ (E_{K^+}) and Blatt (1988), similarly, found outward-rectifying K^+ channels in intact guard cells. Blatt (1988) concluded that activation of the outward rectified K^+ channels follows the voltage difference ($Vm - E_{K^+}$) which, he believes, guarantees the net movement of K^+ out of guard cells irrespective of the extracellular K^+ concentration. Schroeder and Hagiwara (1989) also found that micromolar concentrations of Ca^{2+} inhibited the inward K^+ channels which fits observations of Ca^{2+} inhibiting stomatal opening but does not explain how K^+ is lost from the guard cells of open stomata (but see below).

Parallel to loss of K^+ an efflux of a counterbalancing anion must occur. This could happen via release of anions through anion channels or as the result of a hypothetical H^+/Cl^- symport coming to a halt. Anion channels have been detected in the plasma membrane of guard cells (Schroeder and Hagiwara, 1989; Keller et al., 1989) and their activation is induced by depolarizing potentials and, in one case, stimulated by micromolar Ca^{2+} levels in the cytoplasm (Schroeder and Hagiwara, 1989). It was concluded that the outward anion current would cause depolarization of the plasma membrane with resulting loss of ions: K^+ uptake channels are inactivated and outward Cl^- channels are activated. Also Ca^{2+}-induced stomatal closure can be explained in terms of Ca^{2+} inactivating the inward rectifying K^+ channel and stimulating the outward rectifying Cl^- channel (Schroeder and Hagiwara, 1989).

Release of ions from the vacuole during stomatal closure may also be initiated by collapse of the pmf across the tonoplast due to inactivation of the H^+ pump: hypothetical K^+/H^+ and Cl^-/H^+ antiports would stop and, additionally, polarization of the tonoplast to negative potentials would activate the Ca^{2+} and voltage dependent channel for release of Cl^- and/or malate.

11.5. Characteristics of the tonoplast and plasma membrane ATPases of guard cells

Within this section only information on the proton pumping ATPases of the tonoplast and plasma membranes will be reported. Although an ATPase which transports calcium across the plasma membrane has been detected in higher plant cells nothing is known about the existence of such an ATPase in guard cells. Similarly, it is not known whether a pyrophosphatase exists in the tonoplast of guard cells though it has been detected in the tonoplast of a number of other cell types.

11.5.1. *The proton pumping ATPase of the plasma membrane*

ATPase activity has been measured in crude extracts of GCP using appropriate specific inhibitors (Fricker and Willmer, 1987; 1990a,b) and in partially purified membrane fractions from GCP (Shimazaki and Kondo, 1987; Nejidat et al., 1986). Results from epidermal tissue will be ignored because of the large contamination of other cell types (see Willmer et al., 1987). However, one study (Blum et al., 1988) using epidermal tissue deserves consideration because of the special attention that was given to purification of the guard cells in situ within the epidermis.

Using a linked enzyme assay (Auffret and Hanke, 1981) and crude homogenates of GCP of *Commelina*, a seasonal variation of KmATP and Vmax was obtained by Fricker and Willmer (1987): Km values varied between 0.33 (July–Sept) and 1.12 mM (Oct–Dec) while Vmax varied between 0.84 pmol GCP^{-1} h^{-1} (July–Sept) and 2.21 pmol GCP^{-1} h^{-1} (Oct–Dec). Similar Km and Vmax values have been reported in other studies for the plasma membrane ATPase of guard cells. Nejidat et al. (1986) reported a Vmax value of 4.5 μmol Pi mg $protein^{-1}$ h^{-1} (= 0.9 pmol GCP^{-1} h^{-1} assuming the guard cell protein content is 152 pg, Birkenhead and Willmer, 1986) for *Commelina* while Shimazaki and Kondo (1987) reported a Vmax value of 55 μmol Pi mg $chlorophyll^{-1}$ h^{-1} (= 1.8 pmol GCP^{-1} h^{-1} assuming 3.5 pg chlorophyll per cell, Outlaw et al., 1980) and a Km value of 0.5 mM ATP for *V. faba*. Using epidermal tissue of *V. faba* with isolated guard cells, a higher Vmax was obtained (510 nmol mg $protein^{-2}$ h^{-1} = 6 pmol GGP^{-1} h^{-1}) and a Km of 0.67 mM ATP Blum et al., 1988). pH optima in the different studies were similar varying between 6.4 and 6.8 (see Fig. 11.4).

The enzyme kinetics are of the Michaelis-Menten type, the enzyme requires Mg^{2+} for its activation and all studies except that of Shimazaki and Kondo (1988) indicate that K^+ stimulates the activity by 30–40% (see Fig. 11.4). Vanadate inhibits the activity while NO_3^-, oligomycin and azide have no effect. DES, DCCD and PCMB, as expected, also inhibit activity. Two studies (Nejidat et al., 1986; Shimazaki and Kondo, 1988) also indicate that relatively low concentrations of Ca^{2+} inhibit activity an observation that may be related to the Ca^{2+}-inhibited stomatal opening or it may simply reflect that Ca^{2+} is competing with Mg^{2+} thereby lowering levels of Mg-ATP, the substrate for the ATPase.

Some preliminary work on the isolation and purification of the ATPase has been made. Using a modified method of Chaney and

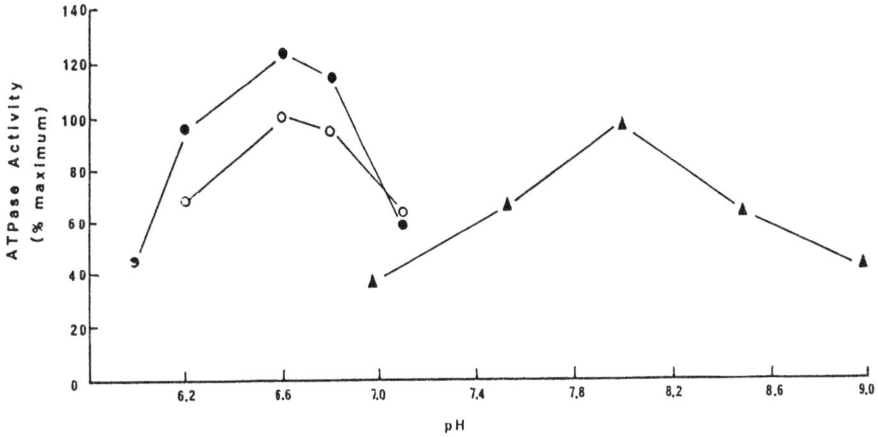

Fig. 11.4. pH profiles for ATPase activities (as a percent of the maximum activity) in crude homogenates of guard cell protoplasts from *Commelina communis*. Key: ●, ○, plasma membrane ATPase activity in the presence and absence of 50 mM KCl, respectively; ▲, tonoplast ATPase activity.

Jacobsen (1983), which involves binding cationic silica microbeads to the outer surface of GCP, Fricker and Willmer (1990b) obtained a plasma membrane-enriched fraction from GCP of *Commelina*. The method is rapid, taking about 10 min to obtain the membrane fraction from the GCP, and the ATPase activity shows a seven-fold increase. On SDS-PAGE gradient gels several polypeptides were enriched in the plasma membrane fraction relative to the remaining fraction, two of which had Mr values slightly above 100 kD, a value similar to the Mr of plasma membrane ATPases in other cell types. Moreover, labelling studies with 14C-DCCD, which covalently binds with ATPase, indicates that a polypeptide with a Mr around 109 kD may be the plasma membrane ATPase of guard cells (Fig. 11.5).

11.5.2. *The proton pumping ATPase of the tonoplast*

The properties of the tonoplast proton pumping ATPase have been investigated less well. Using NO_3^-, inhibition as a diagnostic indicator of tonoplast ATPase activity and crude extracts from GCP of *Commelina*, it has been established that the enzyme is magnesium dependent with a pH optimum of about 8.0 (Fricker and Willmer, 1990b; Willmer and Grammatikopoulos, unpublished data) (Fig 11.4). Enzyme activity is also modulated by light. Thus, alternating light and dark periods increases and decreases activity, respectively, with optimum effect occurring over 15–30 min periods (Fig. 11.6), a time-frame

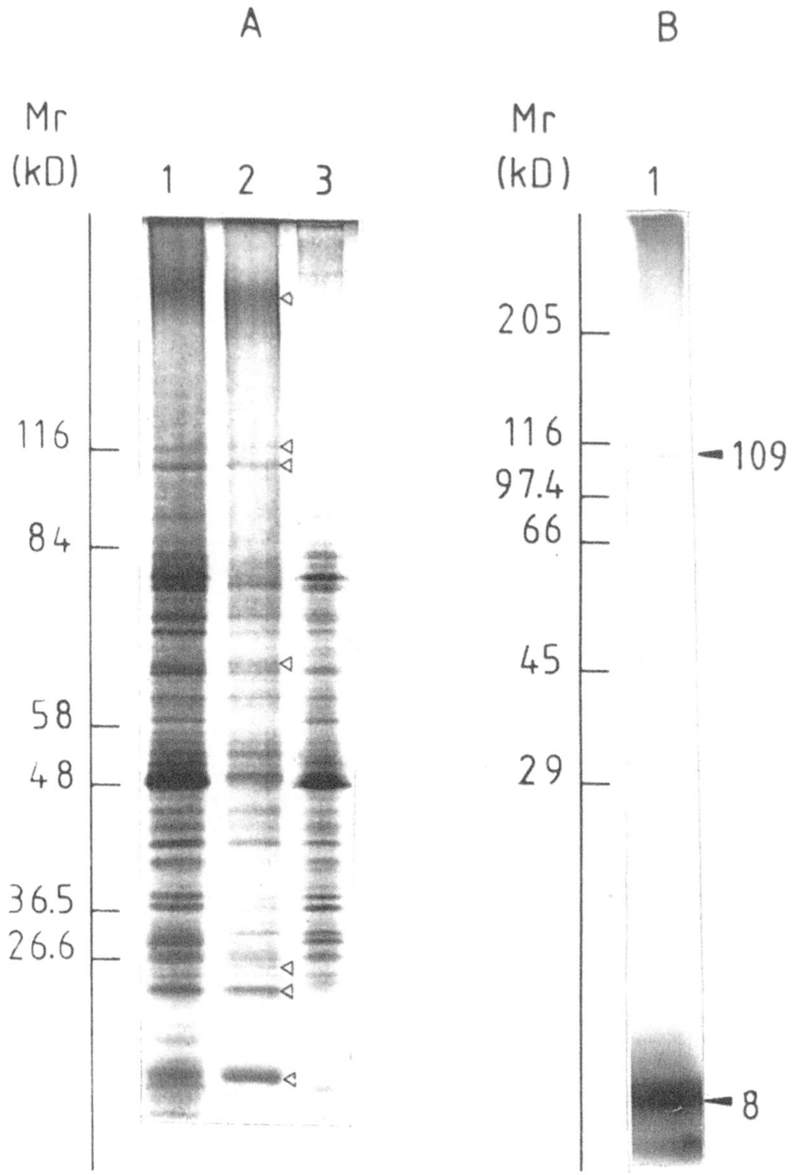

Fig. 11.5. Polypeptide composition of fractions from guard cell protoplasts of *Commelina communis* after separation on SDS-PAGE (5–15% gradient gels) and visualised by silver staining (from Fricker and Willmer, 1990b). *A*: Lane 1 = crude homogenate; Lane 2 = plasma membrane-enriched fraction; Lane 3 = supernatant fraction. The position of marker proteins are indicated. Bands enriched in the plasma membrane fraction are indicated by arrows. *B*: Autoradiography of polypeptides labelled with [^{14}C]DCCD from intact protoplasts after solubilisation and separation by SDS-PAGE. The position of marker proteins are also indicated.

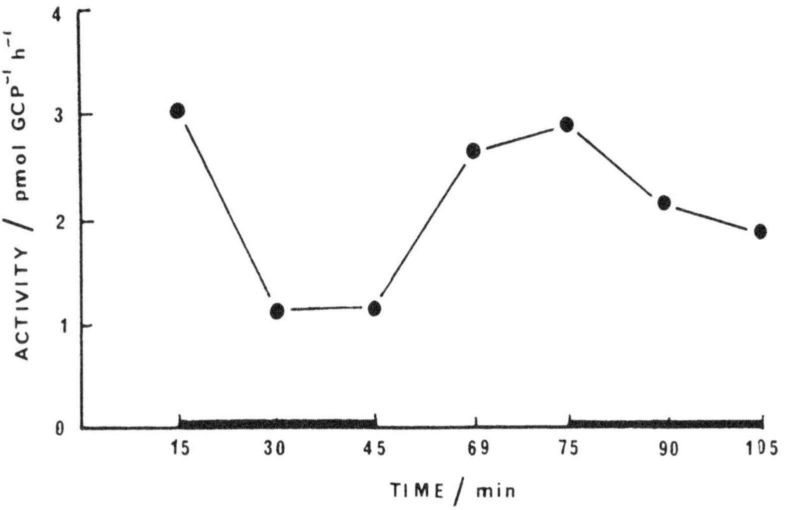

Fig. 11.6. Tonoplast ATPase activity in crude extracts of guard cell protoplasts from *Commelina* in response to periods of light and dark (Grammatikopoulos and Willmer, unpublished).

which fits opening and closing of stomata in vivo (Grammatikopoulos and Willmer, unpublished data).

11.6. Aspects of signal perception and transduction by guard cells

With the finding that specific membrane proteins and secondary messengers are involved in the perception and transduction of stimuli in animal cells a search for similar systems has been made in plant cells. Within plants, guard cells are model systems for such studies because stomata respond rapidly (minutes rather than hours or days) to a variety of environmental and hormonal signals (e.g., light, CO_2 concentration, ABA) by opening or closing their pores. Furthermore, GCP can be used for such studies since they swell and contract in response to hormonal or environmental treatment in an analogous manner to the opening and closing of stomata.

So far, studies on the perception and transduction of signals in guard cells has centred on the mode of action of ABA in bringing about closure of open stomata and preventing closed stomata from opening. Initially ABA binding sites were detected on the outer surface of the plasma membrane of guard cells of *V. faba* (Hornberg and Weiler, 1987) though this has not been confirmed by further investigations. From experiments with epidermal strips of *Commelina* it was concluded that external ABA and Ca^{2+} worked together in a synergistic fashion (De Silva et al., 1985a,b). The latter investigators further hypothesised that ABA opened Ca^{2+} channels in the plasma membrane thereby elevating cytoplasmic Ca^{2+} levels which ultimately brought about closing reactions. Smith and Willmer (1988), however, concluded that external ABA and Ca^{2+} caused GCP of *Commelina* to contract and prevented them from swelling independent of each other. They, therefore, hypothesised that ABA entered the guard cells and was responsible for release of Ca^{2+} from the ER or vacuoles into the cytoplasm which then brought about closing reactions. Recent results from a study by McAinsh et al. (1990) support this view: they found that exogenous ABA induced a rapid increase in free Ca^{2+} within the guard cell cytoplasm which preceded stomatal closure. However, using similar techniques to those used by McAinsh et al. (1990), Gilroy et al. (1991) could not confirm their findings: in about 70% of cases, ABA closed stomata without free Ca^{2+} levels increasing in the cytosol of guard cells.

It has also been suggested that ABA interacts with phospholipids of the plasma membrane of guard cells since it appears to bind specifically to dipalmitoylphosphatidylcholine (DPPC) a lipid found in guard cell membranes (Leshem et al., 1990). Since lipids seem restricted in their ability to transduce signals we contemplate that their may be a protein/phosplolipid association concerned in the binding and transduction of the ABA signal at the plasma membrane.

To confuse the issue further, a number of reports indicate that ABA levels in guard cells increase substantially when leaves are water-stressed (e.g., Harris and Outlaw, 1991). Since it is generally believed that guard cells do not synthesise ABA, at least in significant amounts, it must be transported across the plasma membrane of the guard cells. Thus, it may be that there are no receptor sites for ABA at the outer surface of the plasma membrane but that intracellular ABA levels are the important factor in initiating closure. ABA may bring about release of Ca^{2+} from the ER or the vacuole as suggested by Smith and Willmer (1988) to initiate the machinery for stomatal closure. Alternatively, ABA may be involved in the synthesis of novel proteins which, themselves, are necessary to bring about closing responses. Evidence for this suggestion comes from observations that when GCP of *Commelina* are incubated in 10^{-5} M ABA for 2 h, at least one new polypeptide appears which is absent in control treatments (Willmer, unpublished).

The ABA/Ca^{2+} relationship is further complexed by recent findings that inositol trisphosphate (IP_3), increases free Ca^{2+} in guard cells and stomata close (Blatt et al., 1990; Gilroy et al., 1990). These observations can be used to postulate a mechanism to explain how ABA prevents stomatal opening and brings about closure as follows. ABA binds to the plasma membrane where it stimulates production of IP_3 from phosphatidylinositol-4,5-bisphosphate (PIP_2) (Morse et al., 1989). IP_3 then binds to the tonoplast and evokes release of Ca^{2+} from the vacuole via voltage-dependent channels (Alexandre et al., 1990). The increased cytoplasmic Ca^{2+} finally inactivates the inward rectifying K^+ channels and activates the outward rectifying anion channels in the plasma membrane which results in inducing closure or preventing opening.

Considering the complex responses of stomata to numerous environmental and hormonal signals perhaps it should be expected that ABA is involved in numerous systems, including signal transduction and protein synthesis, which regulate stomatal responses over the short and longer term.

11.7. References and recommended reading

Alexandre, J., Lasalles, J.P. and Kado, R.T. 1990. Opening of Ca^{2+} channels in isolated red beet vacuole membrane by inositol-1,4,5-trisphosphate. *Nature* 343:567–570.

Assmann, S.M., Simoncini, L. and Schroeder, J.I. 1985. Blue light activates electrogenic ion pumping in guard cell protoplasts of *Vicia faba*. *Nature* 318:285–287.

Auffret, C.A. and Hanke, D.E. 1981. Improved preparation and assay and some characteristics of Cl^--ATPase activity from *Limonium vulgare*. *Biochim. Biophys. Acta* 648:186–191.

Birkenhead, K. and Willmer, C.M. 1986. Some biochemical characteristics of guard cell and mesophyll cell protoplasts from *Commelina communis L*. *J. Exp. Bot.* 37:119–128.

Blatt, M.R. 1987. Electrical characteristics of stomatal guard cells: the contribution of ATP-dependent, "electrogenic" transport revealed by current-voltage and difference-current-voltage analysis. *J. Membrane Biol.* 9:257–274.

Blatt, M.R. 1988. Potassium-dependent, bipolar gating of K^+ channels in guard cell. *Planta* 102:235–246.

Blatt, M.R., Thiel, G. and Trentham, D.R. 1990. Reversible inactivation of K^+ channels of *Vicia* stomatal guard cells following the photolysis of caged 1,4,5-trisphosphate. *Nature* 346:766–769.

Blum, W., Key, G. and Weiler, E.W. 1988. ATPase activity in plasmalemma-rich vesicles isolated by aqueous two-phase partitioning from *Vicia faba* mesophyll and epidermis: characterization and influence of abscisic acid and fusicoccin. *Physiol. Plant.* 72:279–287.

Bowling, D.J.F. and Edwards, A. 1984. pH gradients in the stomatal complex of *Tradescantia virginiana*. *J. Exp. Bot.* 35: 1641–1645.

Chaney, L.K. and Jacobson, B.S. 1983. Coating cells with colloidal silica for high yield of plasma membrane sheets and identification of transmembrane proteins. *J. Biol. Chem.* 258: 1062–1072.

De Michelis, M.I., Pugliarello, M.C. and Rasi-Caldogna, F. 1989. Fusicoccin binding to its plasma membrane receptor and the activation of the plasma membrane ATPase. *Plant Physiol.* 90: 133–139.

De Silva, D.L.R., Hetherington, A.M. and Mansfield, T.A. 1985a. Synergism between calcium ions and abscisic acid in preventing stomatal opening. *New Phytol.* 100:473–42.

De Silva, D.L.R., Cox, R.C., Hetherington, A.M. and Mansfield, T.A. 1985b. Suggested involvement of calcium and calmodulin in the responses of stomata to abscisic acid. *New Phytol.* 101:555–63.

Doohan, M.E. and Palevitz, B.A. 1980. Microtubules and coated vesicles in guard-cell protoplasts of *Allium cepa* L. *Planta* 149: 389–401.

Edwards, M.L., Smith, G.N. and Bowling, D.J.F. 1988. Guard cells extrude protons prior to stomatal opening-A study using fluorescence microscopy and pH micro-electrodes. *J. Exp. Bot.* 39:1541–1547.

Fricker, M.D. and Willmer, C.M. 1987. Vanadate-sensitive ATPase and phosphatase activity in guard cell protoplasts of *Commelina*. *J. Exp. Bot.* 38:642–648.

Fricker, M.D. and Willmer, C.M. 1990a. Nitrate-sensitive ATPase activity and proton pumping in guard cell protoplasts of *Commelina*. *J. Exp. Bot.* 41:193–198.

Fricker, M.D. and Willmer, C.M. 1990b. Some properties of proton pumping ATP-ases at the plasma membrane and tonoplast of guard cells. *Biochem. Physiol. Pflanzen.* 186:301–308.

Fujino, M. 1967. Role of adenosinetriphosphate and adenosinetriphosphatase in stomatal movement. *Sci. Bull. Fac. Educ. Nagasaki Univ.* 18:1–47.

Gepstein, S., Jacobs, M. and Taiz, L. 1982/83. Inhibition of stomatal opening in *Vicia faba* epidermal tissue by vanadate and abscisic acid. *Plant Sci. Lett.* 28:63–72.

Gilroy, S., Fricker, M.D., Read, N.D. and Trewavas, A.J. 1991. Signal transduction througn cytosolic calcium in stomatal guard cells of *Commelina communis*. *Plant Cell* 3: 333–344.

Gilroy, S., Read, N.D. and Trewavas, A.J. 1990. Elevation of cytoplasmic calcium by caged calcium or caged inositol trisphosphate initiates stomatal closure. *Nature* 346:769–771.

Gotow, K., Sakaki, T., Kondo, N., Kobiayashi, K. and Syono, K. 1985. Light-induced alkalinization of the suspending medium of guard cell protoplasts from *Vicia faba*. *Plant Physiol.* 79:825–828.

Gradmann, D. and Robinson, D.G. 1989. Does turgor prevent endocytosis in plant cells? *Plant Cell and Environ.* 12:151–154.

Gunar, I.I., Zlotnikova, I.F. and Panichkin, L.A. 1975. Electrophysiological investigation of cells of the stomate complex in spiderwort. *Sov. Plant Physiol.* 22:704–707.

Guyot, M. and Humbert, C. 1970. Les modifications du vacuome des cellules stomatiques d'*Anemia rotundifolia* Schrad. *C.R. Acad Sci.* ser. 270: 2787–2790.

Harris, J.M. and Outlaw, W.H. 1991. Rapid adjustment of guard-cell abscisic acid levels to current leaf water status. *Plant Physiol.* 95:171–173.

Hedrich, R., Barbier-Brygoo, H., Felle, H., Flugge, U.I., Luttge, U., Maathuis, F.J.M., Marx, S., Prins, H.B.A., Raschke, K., Schnabl, H., Schroeder, J.I., Struve, I., Taiz, L. and Ziegler, P. 1988. General mechanisms for solute transport across the tonoplast of plant vacuoles: a patch-clamp survey of ion channels and proton pumps. *Bot. Acta* 101:7–13.

Hornberg, C. and Weiler, E.W. 1987. High-affinity binding sites for abscisic acid on the plasmalemma of *Vicia faba* guard cells. *Nature* 310:321–324.

Imamura, S. 1943. Untersuchungen uber den Mechanismus der Turgorschwankung der Spaltoffnungschlieszellen. *Jap. J. Bot.* 12:251–346.

Inoue, H. and Katoh, Y. 1987. Calcium inhibits ion-stimulated stomatal opening in epidermal strips of *Commelina communis* L. *J. Exp. Bot.* 38:142–149.

Karlsson, P.E. and Schwartz, A. 1988. Characterisation of the effects of metabolic inhibitors, ATPase inhibitors and a potassium channel blocker on stomatal opening and closing in isolated epidermis of *Commelina communis*. *Plant Cell Environ.* 11:165–172.

Keller, B.U., Hedrich, R. and Raschke, K. 1989. Voltage-dependent anion channels in the plasma membrane of guard cells. *Nature* 341:450–453.

Leshem, Y.Y., Cojocaru, M., Margels, S., El-Ani, D. and Landau, E.M. 1990. A biophysical study of abscisic acid interaction with membrane phospholipid components. *New Phytol.* 116:487–498.

McAinsh, M.R., Brownlee, C. and Hetherington, A.M. 1990. Abscisic acid-induced elevation of guard cell cytosolic Ca^{2+} precedes stomatal closure. *Nature* 343:186–188.

Morse, M.J., Satter, R.L., Crain, R.C. and Cote, G.G. 1989. Signal transduction and phosphatidylinositol turnover in plants. *Physiol. Plant.* 76:118–121.

Nejidat, A., Roth-Bejerano, N. and Itai, C. 1986. K, Mg-ATPase activity in guard cells of *Commelina communis*. *Physiol. Plant.* 68: 315–319.

Outlaw, W.H., Mayne, B.C., Zenger, V.E. and Manchester, J. 1980. Presence of both photosystems in guard cells of *Vicia faba* L.: implications for environmental signal processing. *Plant Physiol.* 67:12–16.

Palevitz, B.A. and O'Kane, D.J. 1981. Epifluorescence and video analysis of vacuole motility and development in stomatal cells of *Allium*. *Science* 214:443–445.

Pantoja, O. 1988. Some membrane properties of guard cells. Ph.D. thesis, University of Stirling, Scotland.

Pantoja, O. and Willmer, C.M. 1988. Redox activity and peroxidase activity associated with the plasma membrane of guard cell protoplasts. *Planta* 174:44–50.

Pantoja, O. and Willmer, C.M. 1991. Ferricyanide reduction by guard cell protoplasts. *J. Exp. Bot.* 42:323–329.

Raschke, K., Hedrich, R., Reckmann, U. and Schroeder, J.I. 1988. Exploring biophysical and biochemical components of the osmotic motor that drives stomatal movement. *Bot. Acta* 101:283–294.

Raschke, K. and Humble, G.D. 1973. No uptake of anions required by opening stomata of *Vicia faba*: guard cells release hydrogen ions. *Planta* 115:47–57.

Schnabl, H. and Raschke, K. 1980. Potassium chloride as stomatal osmoticum in *Allium cepa* L., a species devoid of starch in guard cells. *Plant Physiol.* 65, 88–93.

Schroeder, J. 1988. K^+ transport properties of K^+ channels in the plasma membrane of *Vicia faba* guard cells. *J. Gen. Physiol.* 92:667–683.

Schroeder, J.I. and Hagiwara, S. 1989. Cytosolic calcium regulates ion channels in the plasma membrane of *Vicia faba* guard cells. *Nature* 338:427–430.

Schroeder, J.I., Hedrich, R. and Fernandez, J.M. 1984. Potassium selective single

channels in guard cell protoplasts of *Vicia faba*. *Nature* 312:361–362.
Schroeder, J.I., Raschke, K. and Neher, E. 1987. Voltage dependence of K^+ channels in guard-cell protoplasts. *Proc. Natl. Acad. Sci., USA* 84:4108–4112.
Serrano, E.E., Zeiger, E. and Hagiwara, S. 1988. Red light stimulates an electrogenic proton pump in *Vicia* guard cell protoplasts. *Proc Natl. Acad. Sci., USA* 85:436–440.
Shimazaki, K., Iino, M. and Zeiger, E. 1986. Blue-light dependent proton extrusion by guard cell protoplasts of *Vicia faba*. *Nature* 319, 324–326.
Shimazaki, K. and Kondo, N. 1987. Plasma membrane H-ATPase in guard cell protoplasts from *Vicia faba* L. *Plant Cell Physiol.* 28:893–900.
Shimazaki, K. and Zeiger, E. 1987. Red light-dependent CO_2 uptake and oxygen evolution in guard cell protoplast of *Vicia faba*.: evidence for photosynthetic CO_2 fixation. *Plant Physiol.* 84:7–9.
Smith, G.N. and Willmer, C.M. 1988. Effect of calcium and abscisic acid on volume changes of guard cell protoplasts of *Commelina*. *J. Exp. Bot.* 39:1529–1539.
Squire, G.R. and Mansfield, T.A. 1972. A simple method of isolating stomata on detached epidermis by low pH treatment: observations on the importance of subsidiary cells. *New Phytol.* 71:1033–1043.
Squire, G.R. and Mansfield, T.A. 1974. The action of fusicoccin on stomatal guard cells and subsidiary cells. *New Phytol.* 73:433–440.
Willmer, C.M., Jamieson, A. and Birkenhead, K. 1987. Leaf epidermal tissue is unsuitable to use for studying biochemical aspects of stomatal functioning. *Plant Sci.* 52:105–110.
Wolfe, J. and Steponkus, P.L. 1983. Mechanical properties of the plasma membrane of isolated plant protoplasts. *Plant Physiol.* 71:276–285.
Zeiger, E., Moody, N., Hepler, P. and Varela, F. 1977. Light-sensitive membrane potentials in onion guard cells. *Nature* 270:270–271.

12. Signal transduction, Ca^{2+}-triggered membrane glycerolipid turnover and growth/senescence equilibria

"Yond' Calcium has a lean and hungry look"
Shakespeare. Julius Ceasar I.ii. 195
(with due apologies to Ca^{2+}ssius)

This topic initially will be dealt with by presenting an overview later to be followed with detailed descriptions of the various signal transduction components. While emphasizing plant mechanisms an attempt will be made to draw certain analogies to mammalian systems which in several aspects have been investigated in greater detail than in plants.

12.1. Overview

A finely balanced equilibrium between membrane phospholipid catabolism on the one hand, and synthesis and remodelling on the other appears to play a key role in the regulation of plant development and senescence, and depending upon which way the scales tip, the plant may either grow and develop, or cease to grow, mature and age. In the catabolic "*phosphatidyl-linoleyl(-enyl) cascade*", entering of Ca^{2+} into the cytosol triggers the catabolic process by binding to calmodulin and activating phospholipase A_2. The latter proceeds to release linoleic or linolenic acid from the *sn*-2 site of intact phospholipid, thus providing substrate for lipoxygenase. The action of lipoxygenase then generates a series of *oxy*-free radicals, ethylene, endogenous Ca^{2+} ionophores, malondealdehyde and jasmonic acid. These may recycle to the membrane, causing the entry of more Ca^{2+} and induction of a further and identical catabolic cycle. With increased cycling, membranes become progressively senescent and undergo biophysical changes altering microviscosity, bulk lipid phase configurations and transition temperatures. The cascade does not appear to be specific for the phospholipid substrate, and it is envisaged that besides phospholipase A_2, both phospholipase B and lipolytic acylhydrolase could participate in the process.

Membrane remodelling and turnover is a parallel process counteracting the above, and proceeds initially by the same Ca^{2+} and possibly calmodulin-triggering, but leads via phospholipase C action and diacylglycerol formation to protein kinase activation and proton pump recharging. It is speculated that auxin and cytokinin, albeit by different pathways, induce this route, for which membrane phosphotidylinositides may be the preferred membrane-associated phospholipid substrate.

The switch that diverts metabolism cither to the promotive or the senescence pathway at present remains a major enigma still to be solved. This putative switch may well be master-controlled by "senescence genes" which could be turned on, or geared to environmental stimuli; alternatively they may be a function of a preprogrammed genetic longevity factor not necessarily dependent on environment.

12.2. Detail

As in animals, phosphatidylinositide (PI, PIP_2, PIP_3 and possibly even PIP_4) metabolism in concert with Ca^{2+}, calmodulin, phospholipases and lipolytic enzymes plays a pivotal role in regulation of the cell's metabolic and catabolic activity. Work carried out in various laboratories has confirmed a similar but not identical sequence of events which culminate in the trans-membrane transfer of physiological stimuli and hence information (Boss and Massel, 1985; Heim and Wagner, 1986; Sandelius and Sommarin, 1986; Irvine et al., 1989; and the excellent mini-review of Einspahr and Thompson, 1990 with references therein). Of particular interest in the Plant Kingdom is the research of the late Ruth Satter and colleagues on the *pulvinus** of the legume *Samanea saman*. This group has added a further tier of information to the pioneering findings of A.W. Galston's at Yale, to the effect that in addition to the phytochrome-mediated red/far-red response, the afore-mentionend phosphatidylinositides, located in the plasma membrane, also contribute to the transduction of the touch response (Morse et al., 1987, 1989).

Other cases of especial interest are of plant hormone (auxin)-promoted growth (Ettlinger and Lehle, 1988) and of osmo-regulation and

* *Pulvinus* – this a touch-sensitive organ situated at the petiole bases of certain plant species which upon tactile stimulus causes infolding of the leaflets on the compound leaves at whose bases it is located. Pulvini are also found at petiole bases of the "touch-me-not" *Mimosa pudica*, and *Albizzia* spp.

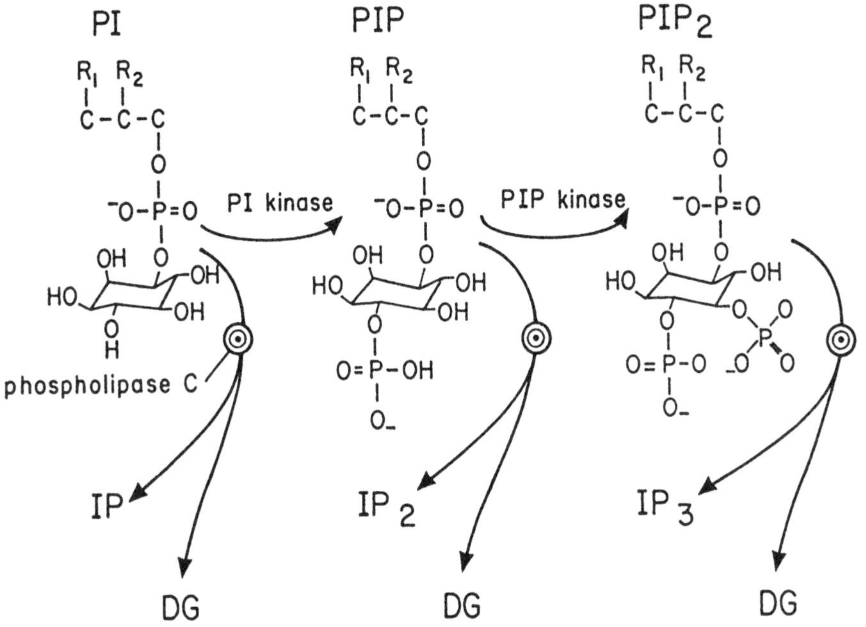

Fig. 12.1. Mediated by a specific kinase at each stage, phosphatidylinositol (PI) is converted to diphosphatidyinositol (PIP) which in turn leads to the formation of triphosphatidylinositol (PIP_2). Note that phospholipase C functioning as a phosphosphodiesterase "severs" the headgroup of these compounds to produce, respectively – mono-, di- and triphosphates of diacylglycerol (DG). (Adapted from Berridge, 1984.)

shock response (Einspahr et al., 1988). However plant membrane scientists are cautioned by Sandelius and Sommarin (1990) not to jump onto the animal phosphatidylinositide signal transduction bandwagon too readily since:

> the available information on polyphosphoinositide metabolism indicates several similarities, but also differences, between plants and animals. PIP phosphorylation and the PIP_2 formed are predominantly localized in the plasma membrane both in plants and animals. The characteristics of PIP kinase in plants and animals are similar, as are those of phospholipase C. One major difference is the relative abundance of PIP and PIP_2. In animal cells, these lipids usually contribute to the same extent to the plasma membrane phospholipids, while both in vivo and in vitro investigations indicate that PIP is more abundant that PIP_2 in plants. Contrary to the situation in animal membranes, PIP_2 is not preferred over PIP as substrate for phospholipase C in plant plasma membranes.

A bird's-eye-view of plant signal transduction in its entirety lends itself to categorization into two parallel and often somewhat contrasting processes:

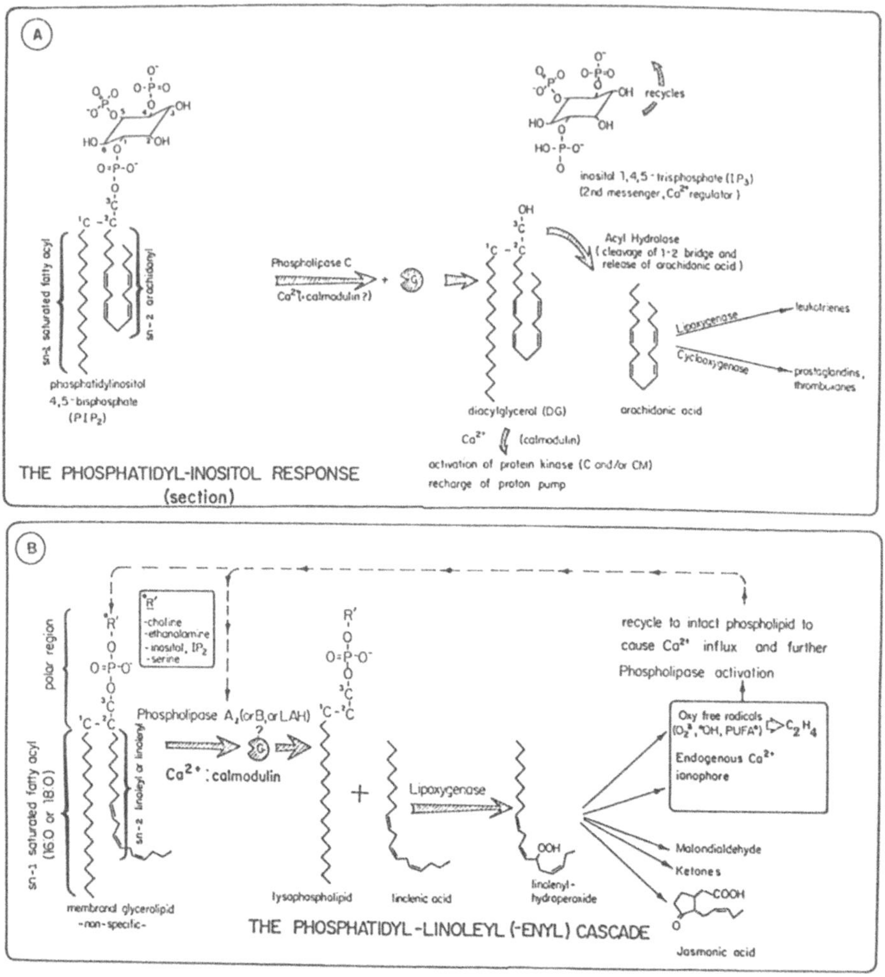

Fig. 12.2.A. Trans-membrane signal transduction, phospholipid metabolism and plant growth regulation.

A. Section of the "PI response" as occurring in mammals and currently investigated in plants as to signal transduction and auxin mode of action (see text). Note the indirect two-step release of the *sn*-2 PUFA chain. The route leading to prostaglandins, etc. – here depicted horizontally – is essentially catabolic, whereas the section leading to release of IP_3, DG and protein kinase C represents functional metabolic turnover.

DG could also be released not directly by phospholipase C as seen above, but also indirectly. This could be achieved by initial action of phospholipase D followed by phosphatidate phosphatase. Note that in most higher plants linoleic or linolenic acid replaces arachidonic acid.

a. *The phosphatidylinositide response*, which is primarily metabolic.
b. *The phosphatidyl-linoleyl(enyl) cascade*, which is catabolic and is associated with cell and organ maturation, senescence and finally death.

In view of the interest generated by the former aspect which as a rule has positive connotations in terms of plant growth regulation, the latter aspect, which may be of no less regulatory importance in plant growth, is often overlooked. The present chapter therefore attempts to address both aspects equally.

Besides the actual detection of phosphatidylinositides in plant membranes which according to Mazliak et al. (1975) comprise ca 15% of

Fig. 12.2.B. The "phosphatidyl-linoleyl(enyl) cascade" as may occur in plant senescence. This scheme indicates a one-step release of *sn*-2 located PUFA and is not specific to PI. It is surmised that senescence results from a shift in the equilibrium from the metabolic route leading to protein kinase activation (see *A* above) to the here-depicted catabolic sequences of events.

In this cascade, entering of Ca^{2+} into the cytosol triggers the catabolic process by activating phospholipase A_2, phospholipase B or lipolytic acyl hydrolase (LAH). The latter, by intervention of "G" protein proceeds to release linoleic or linolenic acid from the *sn*-2 location of intact phospholipid, thus providing substrate for lipoxygenase. The action of lipoxygenase then generates a series of oxy-free radicals – superoxide, hydroxyl and polyunsaturated fatty acid (PUFA) radicals, ethylene, endogenous Ca^{2+} ionophores, malondealdehyde and jasmonic acid. These may recycle to the membrane, causing the entry of more Ca^{2+} and induction of a further, identical catabolic cycle.

With increased cycling, membranes become progressively senescent and undergo biophysical changes altering microviscosity, fluidity, phase configurations of membrane glycerolipids and transition temperatures. The cascade does not appear to be specific for the phospholipid substrate, and can involve all the major categories of membrane-contained glycerolipids including PC, PEA, PS, phosphatidylinositides, as well as galacto-, sulpho-, and possibly even sphingolipids. (This diagram is modified and updated from Leshem (1987) and is reproduced with kind editorial permission from *Physiologia Plantarum*.)

total plasmalemmal glycerolipids and their relative content – (Heim and Wagner (1989) have reported a 40:10:1 ratio between IP, IP_2 and IP_3 in suspensions of cultured *Nicotiana* and *Catharanthus* cells) – the unravelling of their physiological role has been a major step forward towards the establishment of a ubiquitous biological process which as mentioned above is similar but not identical in all biological Kingdoms. A salient contribution is the demonstration in plants of the presence of a series of kinases which are able to convert phosphatidylinositol to -inositides (Sommarin and Sandelius, 1988) as follows:

$$PI \rightarrow PIP \rightarrow PIP_2 \rightarrow PIP_3$$

In the above reactions the direction of the arrows indicates the action of a different type of kinase (see also Fig. 12.1). Following is a description of each of the two transduction processes.

12.3. The phosphatidylinositol response

In mammals this response has been studied extensively and some of the processes it is known to participate in, include the synthesis of the prostaglandin class of hormones, thromboxanes, leucotrienes, immune responses, tumor growth and inhibition, allergies and asthma. Aspirin for example inhibits prostaglandin synthetase and by this alleviates fever, swelling and pain. Moreover most steroidal arthritis-inhibiting drugs are inhibitors of cyclooxygenase. In plants this response has *inter alia* been reported to participate in auxin-stimulated elongation, cell proliferation, flowering, pollen tube elongation, chilling response and acclimation, and stomatal and pulvinar movements. The sequence of events in the phosphatidylinositol response pathway are depicted in Fig. 12.2. The major steps are:

12.3.1. *Ca^{2+} calmodulin binding to phospholipase C*

The initial event is the binding of the activating molecule, termed *agonist* (*agon* in ancient Greek meaning a participant in a sporting event) with a *receptor* molecule. In this particular case the agonist is the divalent Ca^{2+} ion which traverses the membrane from the extrafacial bilayer region (this may also occur, as explained in Chapter 9, from some internal Ca^{2+} store, such as a cell organelle) and enters the cytosol where it combines with calmodulin to form a calcium:calmodulin complex which in turn activates phospholipase C. According to the overall PI-response concept the agonist does not necessarily have to be Ca^{2+} – it may also be a hormone which combines with its specific receptor. Nevertheless in the presently described sequence of events, the Ca^{2+}:calmodulin activation of phospholipase C, Ca^{2+} is believed to fulfill a major agonist function.

The phospholipase C enzyme that has been characterized in plants appears to have a specific substrate requirement – this being solely for PI or for its phosphorylated derivatives (Pfaffman et al., 1987; Melin et al., 1987). Both cytolsolic and particlarate forms of the PI-specific phospholipase C have been found (McMurray and Irvine, 1988) and its presence has been detected in a wide array of plant species belonging to varying plant families.

12.3.2. *Binding to G proteins*

Prior to the phosphodiesterase action of phospholipase C, the next step is the binding of the latter to a *G protein* (see Fig. 12.3). G proteins are proteins that specifically bind to, and are thus activated by guanosine-triphosphate (GTP). The G protein family contains a relatively large number of very closely related members – those that have been characterized are heterotrimeric, with subunits designated α, β and γ in order of decreasing mass.

Probably the most widely studied system in which involvement of a G protein has been implicated, but not rigorously demonstrated, is the regulation of phophoinositide turnover by a phosphatidylinositol-specific phospholipase C (Cockcroft, 1987). The most compelling evidence presented to date involves the capacity of GTP and its non-hydrolyzable analogs to enhance the agonist-mediated accumulation of the products of this enzyme in permeabilized cells and/or membranes derived therefrom. The ability to block this response in some, but not all, cell types by treatment of the cells or membranes with pertussis toxin (which disrupts certain G protein-linked systems) suggests that there may be cell-specific G protein involved. The plethora of information on G protein in mammal (cf review by Casey and Gilman, 1988) has sparked interest in their possible role in the process of plant signal transduction as well.

The plant pattern that is presently emerging is that as in animals, various categories of G proteins are present and by use of antibodies to animal G proteins, cross reactions have suggested that plasma membranes of several plants – *Vicia faba*, *Arabidopsis thaliana*, *Commelina communis* and *Clerodenrum speciosum* – possess immunoreactive G proteins (Blum et al., 1988; Borochov, 1991). Previous reports of a G protein present in crude membrane fractions were provided by Dillenschneider et al. (1986) and Drøback et al. (1988). It however appears that account should be taken of the multiple types of G proteins, some of which not only do not promote GTP incorporation

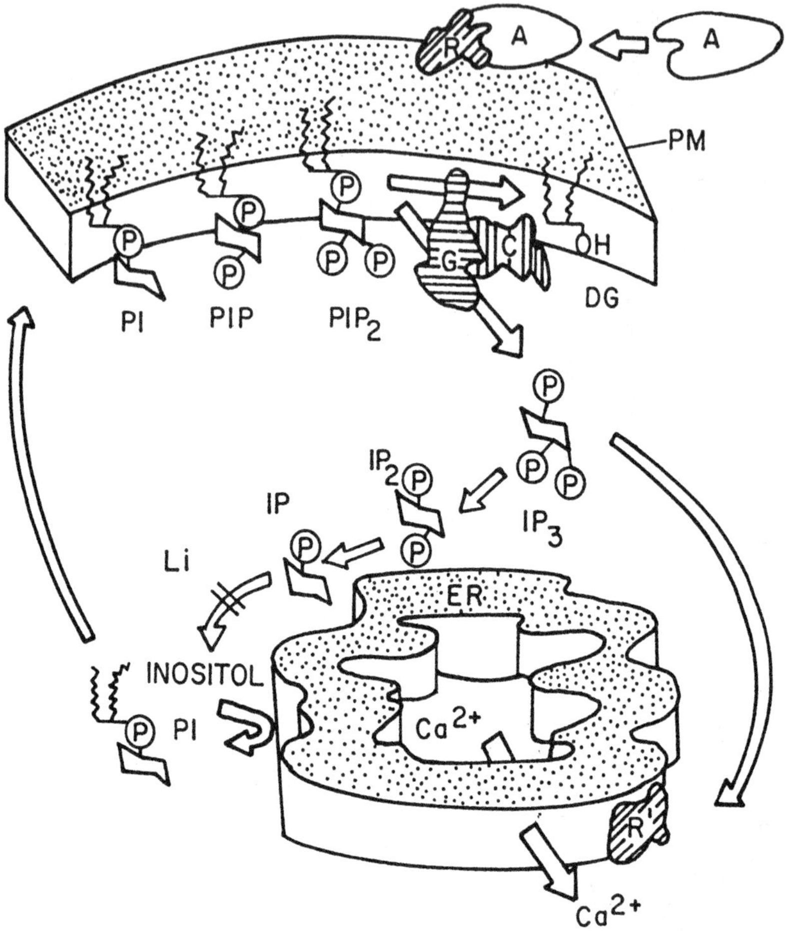

Fig. 12.3. The intracellular sites of the phosphatidylinositol response. In the initial stage the agonist (A) which may be a hormone or Ca^{2+} combines with a receptor (R) located on the outer leaflet of the bilayer comprising the plasma membrane (PM). The A:R complex activates a G protein usually located on the intrafacial side of the membrane. The G protein then binds to phospholipase C which de-esterifies PIP_2 to release IP_3 and DG. The IP_3 combines with another receptor (R′) which is located in the ER, mitochondrial, chloroplast or tonoplast membranes and thereby causes a release of the organelle-stored Ca^{2+}. Concurrently some of the IP_3 may undergo hydrolysis to IP_2 and later to IP and inositol by phosphodiesterase enzymes, i.e. a reverse chain of events to that shown in Fig. 12.1.

The inositol is recycled and serves for biosynthesis of new PI which occurs in the ER. It has been established that the lithium (Li) ion inhibits the enzyme which de-esterifies PI – hence Li prevents resynthesis of IP from inositol. (In the human nervous system lithium slows down responses to external stimuli and the psychiatric use of lithium to moderate manic-depressive situations may possibly be pinpointed to the above-described metabolic stage). The resynthesized IP migrates to the plasma membrane where it is step-wise converted to IP_3 by the various kinase enzymes. (Modified from Morse et al., 1987.)

but rather *inhibit* it, as found in two-phase purified plasmalemmal membranes from soybean hypocotyls (Morré, Chambers and Leshem, unpublished results). Further indication that the G protein stimulation process in plants is not clearly cut and dried is the report that the Ca^{2+} dependent phosphatidylinositidase in soybean membranes has no requirement for guanine nucleotides (Biffin and Hanke, 1990). It thus appears that more work is necessary to determine the degree of resemblance of the plant G protein function and structure to that of animals. An up-to-date discussion of The State of The G Art in plants can be found in the review by Einspahr and Thompson (1990).

12.3.3. *Release of inositoltrisphosphate* (IP_3) *and diacylglycerol (DG)*

As seen in Fig. 12:2 the next step along this pathway is the appearance of IP_3 and DG as a result of action of the phospholipase C/G protein complexing.

IP_3 is believed to be one of the most potent biological chelators of Ca^{2+} and has been included in the list of "secondary messengers" (Berridge and Irvine, 1984). The release of IP_3 is promoted by the hormone indoleacetic acid (Zbell and Walter-Back, 1986, 1988).

The second phospholipase C product, DG, as seen in Fig. 12.2 is a glycerol residue attached to two fatty acyl chains. In both plants and animals DG is an important milestone along the PI pathway in that it serves as an activator of a group of *protein kinase* enzymes which are the topic of the following subsection.

12.3.4. *Protein kinase activation and function*

The protein kinases are a group of pleiotropic enzymes which catalyze the reversible transfer of the γ phosphoryl group from the nucleoside-triphosphates to proteins (Hetherington et al., 1990). Such protein phosphorylation plays a major role in regulation of both enzymic and non-enzymic proteins:

$$\text{protein} + \text{ATP} \xrightarrow[\text{Mg}^{2+} \text{ or Mn}^{2+}]{\text{protein kinases}} \text{protein P} + \text{ADP}$$

The site of phosphorylation on the protein is usually on the serine, threonine or tyrosine residue. In the area of such protein serine/threonine kinases, recently much has been learned about protein kinase C structure and function. Novel lipid mediators, both stimulatory and inhibitory, have been discovered, and kinase has been shown to be an increasingly large family of gene products. Heterogeneity of cellular localization and function has been documented. Calcium/calmodulin-dependent protein kinases are now believed to consist of at least five enzymes, which range from those with extreme substrate specificity such as phosphorylase kinase, to calcium calmodulin kinase II, with *several* known substrates. Several of these enzymes, as for example calcium/calmodulin kinase III, appear to be important in the regulation of protein synthesis (Blackshear et al., 1988). DG-activated protein kinase C possessing a Ca^{2+} (and possibly a calmodulin) requirement has been found in various plant tissues and organs, a typical example of which is the 84.5 KD protein characterized in *Ameranthus tricolor* plants (Elliot et al., 1988). However at least one form of protein kinase, present in soybeans, although Ca^{2+} dependent, has no requirement for calmodulin (Harmon, 1987).

Besides general dependence on DG, Ca^{2+} and sometimes calmodulin, protein kinase C activity is also subject to regulation by other factors including many biophysical ones such as membrane fluidity, and phase configuration, mentioned in Chapter 7 (Epand and Lester, 1990). It is interesting that while DG, containing two fatty acyl groups, activates protein kinase C, certain *single* long tailed fatty acids have been reported to inhibit the enzyme (Polya et al., 1990).

Protein kinase has been reported to be located in the plasmalemma (Klucis and Polya, 1988) of beet cells, this finding not excluding the possibility that it may also appear in a soluble form in the cytosol. As in the case of phospholipase C activity, protein kinase-mediated protein phosphorylation is promoted by auxin (Morré, 1988; Zbell and Walter-Back, 1988). In addition to auxin-induced protein phosphorylation enhancement as previously mentioned, it has been found that cytokinins may act in a like manner (Talkenau et al. 1987). Scherer et al. (1990) have suggested that auxin-promoted PLA_2 producing *lyso*-PLs causes an activation of protein kinase and thereby contributes to signal transduction, furthermore they contend that the rare phospholipid PAF participates in the process.

These reports may be interpreted either as a direct protein kinase C activation by the hormones or alternatively as an indirect one via enhancement of phospholipase C activity which is an earlier event in the chain of events presently being dealt with.

The reader particularly interested in protein phosphorylation in greater detail is referred to the reviews of Ranjeva and Boudet (1987) and of Poovaiah et al. (1987). In addition, Leopold (1990) has critically discussed the issue of plant hormone interaction with the PI response and has clearly distinguished between what is known, what is hypothetical, and what at present remains unknown.

The final outcome of this branch of the PI response – protein phosphorylation – is interpreted as the overall biological activation of many key enzymes and proteins which stimulate physiological activity in processes mentioned at the beginning of this section and in other, as yet non-defined, responses as well. One key mechanism worthwhile of especial mention is the putative protein kinase C charging of the proton pump which according to some workers reflects the basic mode of auxin action in its well-known "acid effect" manifestation.

12.4. The phosphatidyl-linoleyl(enyl) cascade

In contrast to the PI response, the presently described one as said is essentially degradative and thus associated with processes which lead to maturation, senescence and finally death. The sequence of events in the phosphatidyl-linoleyl(enyl) cascade as depicted in Fig. 12.2 is as follows:

12.4.1. *Ca^{2+}-induced phospholipase A_2 activation*

Due to a physiological stimulus or membrane perurbation which causes change in the bilayer surface tension, extraneous Ca^{2+} intrudes into the cytosol – a like cytosolic Ca^{2+} enrichment possibly occurring from organelle-contained Ca^{2+} stores. This Ca^{2+} increment serves to set the cascade into motion by binding to calmodulin. Subsequently the Ca^{2+}:calmodulin complex activates latent phospholipase A_2, B or lypolytic-acylhydrolase (See Fig. 9.1 on page 175).

The common feature of all three of these enzymes is the de-esterification of *sn*-2 contained fatty acyl chain which as mentioned in Chapter 2 is usually a polyunsaturated fatty acid – prevalently 18:2 or 18:3 i.e. linoleic or linolenic acid (hence the above designation of the cascade). A further common characteristic of both the previously-described PI-pathway and of the phosphatidyl-linoleyl cascade is the initial entry of external Ca^{2+} into the cytosol – but the major difference is that in the former the initial enzyme activated is phospholipase C, whereas in the latter it is phospholipase A_2. The released 18:2 or 18:3, now *free* polyunsaturated fatty acids (PUFA's), which in themselves are undesirable since they act as detergents as well, provide substrate for the next stage.

12.4.2. *Hydroperoxidation by lipoxygenase and its immediate consequences*

Lipoxygenase (LOX) preferentially attacks *released* PUFA's – thus their release from intact PL by PLA_2-mediated de-esterification causes an upsurge of LOX activity. This catabolic stage is particularly perilous to cell integrity since LOX activity involves the production of a wide spectrum of danger-fraught free radicals including the PUFA free radical, super-oxide ($O_2^{\cdot}$, the peroxyl and hydroxyl free radicals ($\cdot$ROOH and $\cdot$OH) (see Fig. 12.2). These free radicals may moreover be instrumental in conversion of aminocyclopropane carboxylic acid (ACC) precursor into senescence-evoking ethylene. Furthermore we again mention the earlier contention that free PUFA's may serve as endogenous Ca^{2+} ionophores.

Be this as it may regarding ionophoretic involvement, the implement of free radicals, especially of the *oxy*-free radical species listed above, may cause membrane leaks and causes a breakdown of the outside/inside membrane Ca^{2+} gradient resulting in a massive influx of Ca^{2+} into the cytosol, which in turn activates more phospholipase A_2, which then releases more *sn*-2 contained PUFA, incuding further LOX hydroperoxidation and so on ad mortem. In this manner the cell metaphorically "calcifies itself to death".

12.4.3. *Later events and consequences of membrane glycerolipid peroxidation*

Fig. 12.2 also delineates the pathway of production, and the consequences thereof, of further senescence-evoking compounds. These include *malondealdehyde*, *jasmonic acid* (or its derivates). A glance at the formula of the latter (Fig. 12.2 below) and of prostaglandins (Fig. 12.2, above) indicates the structural if not the functional similarity. *Jasmonic acid*, *methyl jasmonate* and similar "hairpin-bent" products of this pathway including *cucurbic acid* and *tuberonic acid*, are known to promote senscence. In potatoes and Jerusalem artichoke, tuberonic acid moreover induces tuber formation, a process which also may be considered to be senescence-associated (Koda et al., 1991, and see also pages 187–188).

A further biochemical change which is inherently contingent upon the progress of the phosphatidyl-linoleyl cascade is a decrease in the phospholipid/sterol ratio the significance of which has been discussed in the chapter on sterols.

12.4.4. *Biophysical changes in membrane properties associated with the cascade*

The progress of the cascade overtly observed as overall senescence, in terms of membrane biophysics, is accompanied by a bulk lipid phase transition from the fluid-crystalline (L-α) to the gel phase (L-β). Since in this process the phospholipid acyl sidechains become increasingly constricted in their motility and overall membrane rigidification ensues, surface tension increases and concomitantly the membrane's adaptability to withstand pressure changes resulting from micro- or macro-environmental stress is markedly impaired. Leaks are formed and processes are set into motion whereby the cell senesces and finally dies.

In this cascade there is a further important difference to the PI-response in that essentially *all* of the membrane-contained glycero- and sphingolipids may serve as substrate for de-esterification and its consequences, whereas in the PI response only the phosphatidylinositides fulfill this function.

Detailed discussion and documentation of this Ca^{2+}-induced phosphatidyl-linoleyl(enyl) cascade may be found elsewhere (Leshem, 1987). In the capacity of Ca^{2+} to serve as a trigger for the overall agening process in mammals as well, a similar report recapitulating the same sequence of events has recently been presented by McTavish et al. (1990). Working on the rotifer species *Asplanchna brightwelli* they showed that with age, cellular uptake of Ca^{2+} increases. However upon administration of very low concentrations of the Ca^{2+} channel blocker *nifendipine* lifespan was considerably increased. These workers suggest, in keeping with the hypothesis here presented for plant systems, that there is an age-related decline in calcium homeostasis, and support the theory that calcium ion regulation is a vital factor in ageing tissues.

12.5. References and recommended reading

Berridge, M.J. and Irvine, R.F. 1984. Inositoltrisphosphate. A novel second messenger in cellular signal transduction. *Nature* 312:315–21.

Biffin, M. and Hanke, D. E. 1990. Polyphosphoinositdase activity in soybean membranes is Ca^{2+} dependent and shows no requirement for guanine nucleotides. *Plant Science* 69:147–56.

Blackshear, P. J., Naim, A. G. and Kuo, J. F. 1988. Protein kinases 1988: a current perspective. *FASEB Jour.* 2:2957–69.

Blum, W., Hinsch, K.D., Schultz, G. and Weiler, E.W. 1988. Identification of GTP binding proteins in the plasma membrane of higher plants. *Biochem. Biophys. Res. Comm.* 156:54–9.

Borochov, A. 1991. Personal communication.

Boss, W.F. and Massel, M.O. 1985. Polyphosphoinositides are present in plant tissue culture cells. *Biochem. Biophys. Res. Comm.* 132:1018–23.

Casey, P.J. and Gilman, A.G. 1988. G protein involvement in receptor-effector coupling. *Jour. Biol. Chem.* 203:2577–80.

Cockcroft, S. 1987. Polyphosphoinositide phosphodiesterase: regulation by a novel guanine nucleotide binding protein, Gp. *Trends Biochem. Sci.* 12:75–8.

Dillenschneider, M.A., Hetherington, A., Graziana, A., Albert, A., Berta, P., Haiech, J. and Ranjeva, R. 1986. The formation of inositol phosphate derivatives by isolated membranes from *Acer pseudoplatanus* is stimulated by guanine nucleotides. *FEBS Let.* 208:41308.

Drøback, B.K., Allan, J.G., Comerford, J.G., Roberts, K. and Dawson, P. 1988. Presence of guanine nucleotide-binding protein in a plant hypocotyl microsomal fraction. *Biochem. Biophys. Res. Comm.* 150:899–903.

Drøback, B.K. and Ferguson, I.B. 1985. Release of Ca^{2+} from plant hypocotyl microsomes by inositol-1,4,5 trisphosphate. *Biochem. Biophys. Res. Comm.* 130:1241–6.

Einspahr, K.J., Peeler, T.C., Thompson, G.A. Jr. 1988. Rapid changes in polyphosphoinositide metabolism associated with the response of *Dunaliella salina* to hypoosmotic shock. *Jour. Biol. Chem.* 263:5775–9.

Einspahr, K.J. and Thompson, G.A. Jr. 1990. Transmembrane signaling via polyphosphoinositol 4,5-bisphosphate hydrolysis in plants. *Pl. Physiol.* 93:361–6.

Elliott, D.C., Fournier, A. and Kokke, Y.S. 1988. Phosphatidylserine activation of plant protein kinase *C. Phytochem.* 27:3725–30.

Epand, R.M. and Lester, D.S. 1990. The role of membrane biophysical properties in the regulation of protein kinase C activity. *Trends in Pharm. Sci.* 11:317–20.

Ettlinger, C. and Lehle, L. 1988. Auxin induces rapid changes in PI metabolites. *Nature* 331: 176–7.

Harmon, A.C., Putnam-Evans, A. and Cormier, M.J. 1987. A calcium dependent but calmodulin independent protein kinase from soybean. *Plant Physiol.* 83:830–7.

Heim, S. and Wagner, K.G. 1986. Evidence of phosphorylated phosphatidylinositols in the growth cycle of suspension cultured plant cells. *Biochem. Biophys. Res. Comm.* 134:1175–81.

Heim, S. and Wagner, K.G. 1989. Inositol phosphates in the growth cycle of suspension cultured plant cells. *Plant Sci.* 63:159–65.

Hemberg, M., Miersch, O. and Sembdner, G. 1988. Absolute configuration of phytodienoic acid (a jasmonic acid precursor). *Lipids* 23:521–4.

Hetherington, A.M., Battey, N.H. and Millner, P.A. 1990. Protein kinase. pp. 371–83. In: Ed. P.J. Lea. *Methods in Plant Biochemistry*. Academic Press, London and New York.

Irvine, R.F., Letcher, A.J., Lander, D.J., Drøback, B.K., Dawson, A.P. and Musgrave, A. 1989. PI 4,5 bisphosphate and PI-4 phosphate in plant tissues. *Plant Physiol.* 89:882–92.

Klucis, E. and Polya, G.M. 1988. Localization solubilization and characterization of plant membrane associated calcium dependent protein kinases. *Plant Physiol.* 88:164–71.

Koda, Y., Kikuta, Y., Takazaki, H., Tsujino, Y., Sakamura, S. and Yoshihara, T. 1991. Potato tuber inducing activities of jasmonic acid and related compounds. Phytochem. 30: 1435–8.

Leopold, A.C. 1990. Calcium and second messengers in hormonal regulation. pp. 203–8. In: Eds. R.P. Pharis and S.B. Rood. *Plant Growth Substances 1988.* Springer Verlag, Berlin.

Leshem, Y. 1987. Membrane phospholipid catabolism and calcium activity in the control of plant senescence (minireview). *Physiol. Plant.* 69:551–9.

Leshem, Y.Y. 1988. Plant senescence processes and free radicals (review). *Free Rad. Biol. and Med.* 5:39–44.

Leshem, Y.Y., Halevy, A.H. and Frenkel, C. 1986. *Processes and Control of Plant Senescence*. Elsevier, Amsterdam.

Mazliak, P., Douady, D., Demandre, C. and Kader, J.C. 1975. Exchange processes between organelles involved in membrane lipid biosynhesis. pp. 301–18. In: Eds.

T. Galliard and E.J. Mercer. *Recent Advances in the Chemistry and Biochemistry of Plant Lipids*. Academic Press, New York.

McMurray, W.C. and Irvine, R.F. 1988. Phosphatidyl 4,5-bisphosphate phosphodiesterase in higher plants. *Biochem. Jour.* 249:877–81.

McTavish, A., Sawada, M. and Enesco, H.E. 1990. Nifedipine influences rotifer lifespan: studies on the calcium theory of aging. *Age* 13:65–71.

Melin, P.M., Sommarin, M., Sandelius, A.S. and Jergil, B. 1987. Identification of Ca^{2+} stimulated PI phospholipase C in isolated plant plasma membranes. *FEBS Lett.* 223:87–91.

Morré, D.J. 1988. Stimulus-response coupling in auxin regulation of plant cell elongation. pp. 81–114. In: Eds. W. Boss and D.J. Morré. *Second Messengers in Plant Growth and Development*. Alan R. Liss, New York.

Morse, M.J., Crain, R.C. and Satter, R.L. 1987. Phosphatidylinositol cycle metabolites in *Samanea saman pulvini*. *Pl. Phys.* 83:640–4.

Morse, M.J., Satter, R.L., Crain, R.C. and Coté, C.G. 1989. Signal transduction and PI turnover in plants. *Physiol. Plant.* 76:118–21.

Nowack, D.D., Morré, D.M., Paulik, M., Keenan, T.W. and Morré, D.J. 1987. Intercellular membrane flow; reconstitution of transition membrane vesicle formation in a cell free system. *Proc. Nat. Acad. Sci. USA* 84:6098–102.

Pfaffman, H., Hartmann, E., Brightman, A.O. and Morré, D.J. 1987. Phosphatidylinositol specific phospholipase C of plant systems. *Pl. Phys.* 85:1151–5.

Polya, G.M., Minichiello, J., Nott, R., Klucis, E. and Keane, P.J. 1990. Differential inhibition of plant Ca-dependent protein kinases by long chain fatty acids and other amphiphiles. *Plant Sci.* 71:45–54.

Poovaiah, B.W., Reddy, A.S.N. and McFadden, J.J. 1987. Calcium messenger system; role of protein phosphorylation and inositol bisphospholipids (minireview). *Physiol. Plant.* 69:569–73.

Ranjeva, R. and Boudet, A.M. 1987. Phosphorylation of proteins in plants: from metabolic regulation to potential involvement in stimulus response coupling. *Ann. Rev. Pl. Phys.* 38:73–93.

Rincon, M. and Boss, W.F. 1987. Myo-inositol trisphosphate mobilizes calcium from fusogenic carrot protoplasts. *Pl. Physiol.* 83:395–8.

Sandelius, A.S. and Morré, D.J. 1987. Characteristics of a phosphatidylinositol exchange activity of soybean microsomes. *Pl. Phys.* 84:1022–7.

Sandelius, A.S. and Sommarin, M. 1986. Phosphorylation of phosphatidylinositol in isolated plant membranes. *FEBS Lett.* 201:282–6.

Sandelius, A.S. and Sommarin, M. 1990. Membrane-localized reactions involved in polyphosphoinositide turnover in plants. In: Eds. D.J. Morré, W.F. Boss and F. Loewas. *Inositol Metabolism in Plants*. Alan R. Liss, New York (In press).

Scherer, G.F.E., André, B. and Martiny-Baron, G. 1989. A rapid response to a plant hormone: auxin stimulates phospholipase A_2 in vivo and in vitro. *Biochem. Biophys. Res. Comm.* 163:111–17.

Scherer, G.F.E., André, B. and Martiny-Baron, G. 1990. Hormone activated phospholipase A_2 and lysophospholipid activated protein kinase: a new signal s transduction chain: a new, second messenger system in plants? *Curr. Top. Plant. Biochem., Physiol.* 9 (In press).

Sommarin, M. and Sandelius, A.S. 1988. Phosphatidylinositol and phosphatidylinositolphosphate kinases in plant plasma membranes. *Biochim. Biophys. Acta* 958:268–78.

Talkenau, E., Heim, S. and Wagner. 1987. Effect of cytokinins on the phospholipid phosphorylation of the suspension cultured *Catharanthus roseus* cells. *Pl. Sci.* 50:173–8.

Zbell, B. and Walter-Back, C. 1986. About the search for the molecular action of high affinity auxin-binding sites on membrane localized rapid PI metabolism in plant cells. pp. 141–54. In: Ed. D. Klämbt. *Plant Hormone Receptors*. Springer Verlag, Berlin.

Zbell, B. and Walter-Back, C. 1988. Signal transduction of auxin in isolated plant cell membranes; indications for a rapid polyphosphoinositide response stimulated by indoleacetic acid. *Jour. Pl. Phys.* 133:353–60.

Epilogue

Not to terminate this chapter and book on a senescence-sounding note, it should be remembered that in plants, contrary to man, ageing does not as of necessity have negative biological connotations. It is sometimes of considerable agricultural and economical value to hasten fruit or vegetable maturation – which is an ageing process – in order to ensure supply of produce during holiday seasons when market demand and prices are high. Moreover, localized senescence of certain plant organs is accompanied by hydrolysis of carbohydrate, protein and lipid reserves and triggers transport to other organs – fruit and seed (including cereal caryopses) – to ensure perpetuation of the species or, in horticultural terms, maximal yield.

In shrub and perennial deciduous tree species the degree of floral bud awakening (and hence yield) in the spring, usually preceding the awakening of photosynthesizing vegetative buds, may be a direct function of availability of these shoot- or root-contained reserves which in turn are the products of senescence processes taking place in foliage in the previous fall. It thus follows that foliar cell membrane disruption and lysis prior to leaf abscission may be a prerequisite to ensure optimal and unhindered diffusion of cytosol-contained hydrolysates to vascular conduits connecting leaf tissue to storage organs. On the other hand under certain conditions senescence deferral for the purpose of long distance marketing of fresh produce as in the case of flowers and fruit, or in order to regulate work schedules in farms, may be as advantageous if not more so than its promotion.

As outlined in the various chapters of this book, it now appears to be increasingly evident that plant membranes play an important role in the doubtless many-faceted plant growth, stress and senescence phenomena, the expression and interpretation of which require a multi-disciplinary approach including aspects of classical botany, molecular biology, plant physiology, biochemistry and biophysics. To these, the ingredients of time and patience to contemplate and to attempt to integrate the various disciplines in a logical and coherent manner, should be added. If this book serves as an aid towards the achieving of this elusive aim, it will well have served its purpose.

Abbreviations

ABA–	abscisic acid
AB-In–	antibody immunoglobulin
ACC–	1-aminocyclopropane-1-carboxylic acid
CL–	cardiolipin
CM–	calmodulin
c PnA–	*cis*-parinaric acid
2,4-d–	2,4-dichlorophenoxyacetic acid
DAG or DG–	diacylglycerol
DCCD–	N,N dicyclohexyl carbodiimide
DES–	diethylsilbestrol
DGDG–	digalactosyldiacylglyceride
D_L–	lateral (translational) diffusion
DPD–	diphosphatidylglycerol
DPH–	docosapentaenoic acid
DPHX–	diphenylhexatriene
DPPC–	dipalmitoylphosphatidylcholine
DSC–	differential scanning colorimetry
D_r–	rotational diffusion
D_T–	translational (= lateral) diffusion
EDTA–	ethylenediaminetera-acetic acid
EPA–	eicosapentaenoic acid
EPR–	electron paramagnetic resonance
ESR–	electron spin resonance
FA–	fatty acid
FC–	fusicoccin
FRAP–	fluorescence recovery after photobleaching
G–	guanosine
GA–	gibberellic acid
GCP–	guard cell protoplasts
GlcCer–	glucosylceramide
gp–	glutathione peroxidase

GSH–	glutathione
GTP–	guanosine-triphosphate
HDL–	high density lipoprotein
HPLC–	high pressure liquid chromatography
I–	inositol
IAA–	indole-3-acetic acid
IBA–	indole-3-butyric acid
IDL–	intermediary density lipoproteins
in-a Z–	ice nucleating protein
IP–	inositol 1-phosphate
IP_2–	inositol 1,4-bisphosphate
IP_3–	inositol 1,4,5-tris-phosphate
IPMS–	infinite periodic minimal surface
LAH–	lipolytic-acylhydrolase
LHC II–	light harvesting complex II
LOX–	lipoxygenase
LTP–	lipid transfer proteins
LUV–	large unilamellar vesicles
MCPA–	methylchlorophenoxiacetic acid
MDA–	malondealdehyde
MGDG–	monogalactosyldiacylglyceride
MLV–	multivesicular large vesicles
mN–	milliNewtons
MPD–	monophosphatidylglycerol
NAA–	naphthaleneacetic acid
NAD–	nicotine-adenine-dinucleotide
NPA–	naphtylphtalamic acid
PA–	phosphatidic acid
PAF–	platelet activating factor
PC–	phosphatidylcholine (lecithin)
PCMB–	para chloromercuribenzoic acid
PE–	phosphatidylethanol
PEA–	phosphatidylethanolamine
PFD–	proton flux density
PG–	phosphatcolylglycerol
PHP–	phosphatidatephosphatase
PI–	phosphatidylinositol
PIP–	phosphatidylinositol 4-phosphate
PIP_2–	phosphatidylinositol 4,5-bisphosphate
PIP_3–	phosphatidylinositol 1,4,5-trisphosphate
PIP_4–	phosphatidylinositol 1,3,4,5-tetrakisphosphate
PL–	phospholipid

PLA_1–	phospholipase A_1
PLA_2–	phospholipase A_2
PLB–	prolammelar body
PL-B–	phospholipase B
PL-C–	phospholipase C
PL-D–	phospholipase D
PLPC–	palmitoyllinoleylphosphatidylcholine
pmf–	proton motive force
PnA–	parinaric acid
PS–	phosphatidylserine
PS-I–	photosystem I
PUFA–	polyunsaturated fatty acid
SDS-PAGE–	sodium dodecylsulphate-polyacrylamide gel electrophoresis
sn–	specific numeration
SpL–	sphingolipid
SQDG–	sulphoquinovosyldiacylglycerol
SUV–	small unilammelar vesicles
T_c–	transition temperature
TFP–	trifluoperazine
TIBA–	tri-iodobenzoic acid
TMA-DPHX–	trimethylamine diphenylhexatriene
*t*PnA–	*trans*-parinaric acid
UFA–	unsaturated fatty acid
VLDL–	very low density lipoprotein

Index

GPSR Compliance

The European Union's (EU) General Product Safety Regulation (GPSR) is a set of rules that requires consumer products to be safe and our obligations to ensure this.

If you have any concerns about our products, you can contact us on ProductSafety@springernature.com

In case Publisher is established outside the EU, the EU authorized representative is:

Springer Nature Customer Service Center GmbH
Europaplatz 3
69115 Heidelberg, Germany

Batch number: 10799857

Printed by Printforce, the Netherlands